Copernicus Books
Sparking Curiosity and Explaining the World

Drawing inspiration from their Renaissance namesake, Copernicus books revolve around scientific curiosity and discovery. Authored by experts from around the world, our books strive to break down barriers and make scientific knowledge more accessible to the public, tackling modern concepts and technologies in a nontechnical and engaging way. Copernicus books are always written with the lay reader in mind, offering introductory forays into different fields to show how the world of science is transforming our daily lives. From astronomy to medicine, business to biology, you will find herein an enriching collection of literature that answers your questions and inspires you to ask even more.

Alfio Quarteroni
Artificial Intelligence
Unveiling the Truth beyond the Myths

Alfio Quarteroni
Politecnico di Milano
Milan, Italy

ISSN 2731-8982 ISSN 2731-8990 (electronic)
Copernicus Books
ISBN 978-3-031-92972-4 ISBN 978-3-031-92973-1 (eBook)
https://doi.org/10.1007/978-3-031-92973-1

Translation from the Italian language edition: "L'intelligenza creata" by Alfio Quarteroni, © Hoepli 2025. Published by Hoepli. All Rights Reserved.

© The Editor(s) (if applicable) and The Author(s), under exclusive license to Springer Nature Switzerland AG 2025

This work is subject to copyright. All rights are solely and exclusively licensed by the Publisher, whether the whole or part of the material is concerned, specifically the rights of reprinting, reuse of illustrations, recitation, broadcasting, reproduction on microfilms or in any other physical way, and transmission or information storage and retrieval, electronic adaptation, computer software, or by similar or dissimilar methodology now known or hereafter developed.
The use of general descriptive names, registered names, trademarks, service marks, etc. in this publication does not imply, even in the absence of a specific statement, that such names are exempt from the relevant protective laws and regulations and therefore free for general use.
The publisher, the authors and the editors are safe to assume that the advice and information in this book are believed to be true and accurate at the date of publication. Neither the publisher nor the authors or the editors give a warranty, expressed or implied, with respect to the material contained herein or for any errors or omissions that may have been made. The publisher remains neutral with regard to jurisdictional claims in published maps and institutional affiliations.

This Springer imprint is published by the registered company Springer Nature Switzerland AG
The registered company address is: Gewerbestrasse 11, 6330 Cham, Switzerland

If disposing of this product, please recycle the paper.

Alfio Quarteroni
Artificial Intelligence
Unveiling the Truth beyond the Myths

Alfio Quarteroni
Politecnico di Milano
Milan, Italy

ISSN 2731-8982   ISSN 2731-8990 (electronic)
Copernicus Books
ISBN 978-3-031-92972-4   ISBN 978-3-031-92973-1 (eBook)
https://doi.org/10.1007/978-3-031-92973-1

Translation from the Italian language edition: "L'intelligenza creata" by Alfio Quarteroni, © Hoepli 2025. Published by Hoepli. All Rights Reserved.

© The Editor(s) (if applicable) and The Author(s), under exclusive license to Springer Nature Switzerland AG 2025

This work is subject to copyright. All rights are solely and exclusively licensed by the Publisher, whether the whole or part of the material is concerned, specifically the rights of reprinting, reuse of illustrations, recitation, broadcasting, reproduction on microfilms or in any other physical way, and transmission or information storage and retrieval, electronic adaptation, computer software, or by similar or dissimilar methodology now known or hereafter developed.
The use of general descriptive names, registered names, trademarks, service marks, etc. in this publication does not imply, even in the absence of a specific statement, that such names are exempt from the relevant protective laws and regulations and therefore free for general use.
The publisher, the authors and the editors are safe to assume that the advice and information in this book are believed to be true and accurate at the date of publication. Neither the publisher nor the authors or the editors give a warranty, expressed or implied, with respect to the material contained herein or for any errors or omissions that may have been made. The publisher remains neutral with regard to jurisdictional claims in published maps and institutional affiliations.

This Springer imprint is published by the registered company Springer Nature Switzerland AG
The registered company address is: Gewerbestrasse 11, 6330 Cham, Switzerland

If disposing of this product, please recycle the paper.

To Lara, Luca, Bianca Sofia, and Leonardo Alexander, so that they may never fear their future

Preface

The issue of artificial intelligence (AI) and its impact on various aspects of human life is a topic that requires deep reflection. In this book, I will seek to explore it in a clear and precise manner.

When we talk about artificial intelligence, we cannot help but ask a question that I am often confronted with during public discussions: "*But is it really intelligent, like human intelligence?*" This is a natural question, which in turn raises many others: "*If AI were truly intelligent, could it ever surpass human capabilities? Could there come a point of no return, where humanity becomes subjugated by these machines? What will happen to jobs in a world dominated by AI? And how can we protect our privacy and data from indiscriminate use?*"

I will attempt to answer these questions while also explaining why, for some of them, there may not be a definitive or exhaustive answer. My goal is to clarify what artificial intelligence truly is, dispelling myths and misconceptions while providing a rigorous and unambiguous perspective on this technological revolution. I will outline its vast potential, the successes already achieved, the hopes it inspires, but also the risks it poses—not only for us as individuals but also for society as a whole.

The intention is not to chase the latest novelty in generative AI or the newest deep neural network architecture. Rather, I will focus on the fundamental concepts that underpin AI, concepts that are meant to endure despite the rapid pace of innovation. I will also highlight the connection between generative AI and other forms of AI, clarifying that not all applications using algorithms can be labeled as "artificial intelligence."

Since the 1950s, pioneers like Alan Turing and John McCarthy have laid the foundations of what we now call AI. Since then, the very definition of AI has evolved continuously, reflecting the variety of goals it has taken on: from the so-called weak or narrow AI, designed for specific tasks, to general AI, which aims to replicate human intelligence, and even the concept of superintelligence, which, in theory, could surpass human capabilities. At present, we remain within the boundaries of narrow AI, which assists humans in specific tasks without truly competing with human intelligence in a broader sense.

Nevertheless, AI possesses extraordinary abilities that we humans struggle to replicate: the capacity to process vast amounts of data, rapidly synthesize complex information, predict consumer behavior with remarkable accuracy, and, more recently, generate texts, poems, stories, images, and films in mere fractions of a second. In healthcare, neural networks can detect diagnoses that might escape even the most skilled specialists. In many cases, AI challenges our understanding of human agency and decision-making, raising questions about control and responsibility.

At the core of AI's progress lies the ability of computers to learn autonomously, primarily—albeit not exclusively—through artificial neural networks. This process relies more on training these networks using available data—often, but not always, big data—than on traditional scientific theory. Naturally, the training data must be relevant to the specific task at hand. In a sense, the experience encoded in data has taken precedence over theory, disrupting the paradigm of modern science.

I will explore the areas where AI has achieved the most remarkable results, offering insights into its future impact and the challenges it presents, such as the enormous resource consumption required to train deep neural networks. This is an issue we cannot ignore, as AI's carbon footprint is rising sharply, posing serious concerns about the environmental sustainability of its development.

Another socially significant issue is AI's impact on the job market. While it will eliminate many professions, particularly those involving repetitive tasks and complex data processing, it will also create new ones. However, this transition will not be without difficulties, and many people may find themselves excluded from this transformation. For this reason, proactive policies in education and training will be essential to ensure that the benefits of AI are distributed as equitably as possible.

AI represents a monumental shift in how we interact with the world and with one another. In many tasks, it far surpasses human capabilities; however, its processes are often opaque, creating the so-called black box effect—where even its creators cannot always explain how an AI system arrives at certain decisions. This phenomenon forces us to reconsider fundamental concepts such as responsibility and trust, especially when AI is applied in critical fields like healthcare, justice, or warfare.

Another key issue is AI's impact on governance and geopolitics. Today, artificial intelligence is taking on a strategic role in global power relations, much like nuclear technology did in the twentieth century. Nations at the forefront of AI development could gain disproportionate influence through economic dominance, military superiority, and data control. This scenario raises crucial concerns about new forms of global inequality and the potential for a technology-driven arms race. AI has the potential to reshape traditional power dynamics, making it essential for governments to develop forward-thinking policies for its regulation and integration into society.

Artificial intelligence is already being used to optimize public services, prevent crime, and even influence electoral behavior through targeted algorithms. However, its use in governance raises serious concerns about privacy and individual freedoms. Without careful oversight, AI could become a tool for mass surveillance or public opinion manipulation, threatening the very foundations of democracy.

Humans are naturally drawn to innovation, yet at the same time, they fear what they do not fully understand or cannot control. It is only natural to be wary of the unknown. The very term *artificial intelligence* evokes a challenge. Regardless of the emotions it stirs in us, AI is becoming omnipresent and is likely to trigger a profound transformation across society, the economy, and politics.

In this book, my aim is neither to glorify nor to demonize AI. Instead, I seek to examine its implications as long as they remain within the realm of human understanding. My goal is to provide the tools to better grasp this phenomenon, fostering a critical and informed approach to the transformation we are experiencing. I also hope to imagine a future that goes beyond AI—or rather, one in which AI is evermore seamlessly integrated with the scientific knowledge that has enabled humanity to achieve the extraordinary progress of the past century.

Milan, Italy Alfio Quarteroni
July 15, 2025

Acknowledgments

I would like to thank Francesca Bonadei, Paola Gervasio, David Moscato, Silvia Quarteroni, and Francesco Regazzoni for their advice on a preliminary version of this text. Paola Gervasio is also sincerely thanked for creating the illustrations in this book. Part of this book was written while the author was in residence at the Oden Institute for Computational Engineering and Sciences at the University of Texas at Austin.

Competing Interests The author has no competing interests to declare that are relevant to the content of this manuscript.

Contents

1 The Early Pages of AI: Between Dream and Reality 1

2 Shining Light on AI 11

3 Machine Learning: When Computers Learn (On Their Own!) . . . 19

4 Mathematics…the Bare Minimum 27

5 What Data for Learning, and What Kind of Learning? 41

6 Generative AI: The Sudden Boom 53

7 AI Is Not Just Generative 71

8 AI in Science 75

9 Where Do We Stand and a Look into the Future 87

10 Artificial Intelligence and Human Intelligence 107

11	Black, White, and Grey Boxes: A New Frontier, from Big Data to Big Science	119
12	Dialoguing with ChatGPT4	131
13	Conclusion and Further Insights	135
14	Addendum: A Brief Mathematical Digression on Complexity, Reproducibility, Interpretability, and Explainability of AI	147

Glossary 151

References 155

1

The Early Pages of AI: Between Dream and Reality

In the first half of the twentieth century, science fiction introduced the concept of artificially intelligent robots to the public. It all began with the humanoid robot that portrayed Maria in Fritz Lang's *Metropolis* in 1927 and continued with the "heartless heart" of the Tin Man in *The Wizard of Oz*, directed by Victor Fleming in 1939. By the 1950s, a generation of scientists, mathematicians, and philosophers had culturally assimilated the concept of "intelligent machines". Among them was Alan Turing, a brilliant British thinker who, in addition to his groundbreaking work on *Enigma*, the machine that enabled the decryption of German messages during World War II (Copeland 2024), was a true pioneer in exploring the mathematical foundations of artificial intelligence. Turing proposed a fundamental question: if humans use information and reason to solve problems and make decisions, why couldn't machines do the same? This logical framework formed the basis of his 1950 paper, *Computing Machinery and Intelligence*, published in *Mind* (Turing, 1950). In this article, Turing famously asked: Can machines think? However, he noted that before answering, it was first necessary to define precisely what is meant by *machine* and *thinking*.

Turing discussed how to build intelligent machines and how to test their intelligence, introducing what would later become the famous *Turing Test*: *A computer is placed in one room, a person in another, and a third participant, unaware of which is which, attempts to determine—through a series of questions—whether they are interacting with a human or a machine. The machine, of course, has the ability to "deceive" the evaluator.* The idea is that if the algorithms with which the machine is "trained" are powerful enough, it can become intelligent to the point of *disguising itself* as a human, successfully fooling the evaluator. As Turing put it:

> *Intelligence is demonstrated by a machine's ability to engage in natural language conversation and deceive a human into believing it is human.*

In doing so, Turing bypassed centuries of philosophical debate on the nature of intelligence. The "imitation game" he introduced suggested that if a machine could operate so effectively that observers were unable to distinguish its behavior from that of a human, then it should be labeled as intelligent. However, Turing was ahead of his time and was unable to turn his insights into practical applications. At that time, computers lacked a fundamental prerequisite for intelligence: they could execute commands but not store them. They could be instructed on *what to do*, but they had no memory of *what they had done*. Additionally, the extremely high computation times and costs posed another major obstacle. These and other limitations led Hans Moravec, then a PhD student under John McCarthy (who, as we will soon see, coined the first definition of AI), to state that "computers were still millions of times too weak to exhibit intelligence." In the early 1950s, the cost of renting a computer could reach $200,000 per month, making them accessible only to the wealthiest and most prestigious universities and major American technology companies. It became essential to convince funding agencies that research into machine intelligence was worth supporting. A few years later, Allen Newell, Herbert A. Simon, and Cliff Shaw developed *Logic Theorist*, a software program deliberately designed to perform automated reasoning. *Logic Theorist*, now considered the first artificial intelligence program, was capable of proving 38 of the first 52 theorems in the second chapter of *Principia Mathematica* by Alfred N. Whitehead and Bertrand Russell, even discovering new and shorter proofs for some of them. The program aimed to replicate human problem-solving abilities, which led to its funding by the *Research and Development (RAND) Corporation*.

A Constantly Evolving Definition

In 1955, when Newell and Simon began working on *Logic Theorist*, the field of artificial intelligence did not yet exist. Even the term *artificial intelligence* itself had not been coined and would not emerge until the following summer. The term was introduced during a historic conference—the *Dartmouth Summer Research Project on Artificial Intelligence*—held at Dartmouth College in Hanover, New Hampshire, in August 1956. At this conference, John McCarthy, Marvin Minsky, Nathaniel Rochester, and Claude Shannon brought together leading researchers from various disciplines for an open discussion on artificial intelligence, a term that McCarthy specifically coined for the occasion. In his proposal, McCarthy, a 28-year-old mathematics professor at Dartmouth College, stated that the conference should *"proceed on the assumption that every aspect of learning or any other feature of intelligence can, in principle, be so precisely described that a machine can be made to simulate it."* He also put forward the following definition of AI:

> *The science and engineering of making intelligent machines, especially intelligent computer programs.*

It is worth noting how, much like Turing before him, McCarthy associated the concept of intelligence with that of machines (or computers). Until recently, the prevailing belief had been that machines—especially computers—could not possess intelligence and that only humans were capable of pursuing goals autonomously and making decisions.

Despite its failure to establish standard methods for the new field, all participants at the conference enthusiastically agreed that AI was achievable. In fact, the Dartmouth College event served as a catalyst for the next 20 years of AI research. From this premise, it becomes clear that the boundaries of AI are not indelibly drawn. They are subjective—since perceptions of what constitutes a uniquely human trait vary from person to person—and they shift over time as new milestones are reached. This phenomenon, known as the *AI effect*, was succinctly captured in 1970 by American computer scientist Larry Tesler, who stated:

> *AI is whatever hasn't been done yet.*

Many other definitions have since been proposed, including those from public institutions and cultural organizations. However, the insights of Turing and McCarthy have remained foundational, redirecting the focus of AI definitions toward performance—behaviors that appear intelligent—rather than the deeper philosophical, cognitive, or neuroscientific aspects of intelligence. For example, Marvin Minsky, one of AI's pioneers, defined it in 1985 as:

> *AI is the science of making machines do things that would require intelligence if done by men.*

According to MIT (Massachusetts Institute of Technology):

> *AI is the quest to build machines that can think, act, and learn like humans.*

This definition embraces a broader concept of cognition and learning rather than referring to specific objectives. In 2019, the European Commission proposed the following definition:

> *AI refers to systems that display intelligent behavior by analyzing their environment and taking actions—with some degree of autonomy—to achieve specific goals.*

Here, the emphasis is placed on autonomy in pursuing specific objectives. According to the OECD (Organization for Economic Co-operation and Development):

> *AI is a machine-based system that is capable of influencing the environment by producing an output (predictions, recommendations, or decisions) for a given set of objectives. It uses machine and/or human-based data and inputs to: (i) perceive real and/or virtual environments; (ii) abstract these perceptions into models through analysis in an automated manner (e.g., with machine learning) or manually; and (iii) use model inference to formulate options for outcomes. AI systems are designed to operate with varying levels of autonomy.*

The Oxford English Dictionary defines AI as:

> *The capacity of computers or other machines to exhibit or simulate intelligent behavior.*

This definition focuses on the ability of machines to mimic behaviors typically associated with human intelligence. As expected since that distant summer of 1956, the definition of AI has evolved over the years but has largely remained within a framework that emphasizes the simulation of human abilities, rational behavior, autonomous learning, and decision-making. To this day, there is no universally accepted definition. However, according to the most widely held perspective:

> *AI encompasses the abilities demonstrated by hardware and software systems that, to a human observer, may appear to be exclusive to human intelligence.*

Winters and Springs

But let us momentarily turn the clock back to that year of grace, 1956. From 1957 to 1974, AI flourished. Computers were able to store more information and became faster, cheaper, and more accessible. Machine Learning algorithms, which are the foundation of Machine Learning (ML), or the ability of computers to learn on their own, will be discussed in Chap. 5 and will accompany us throughout the book. These algorithms have become increasingly effective, and people have become more skilled at knowing which algorithm to apply to their problem. Among the first demonstrations, we can mention the *General Problem Solver (G.P.S.)*, a computer program created in 1957 by H. A. Simon, J. C. Shaw, and Allen Newell, aimed at solving general problems (as long as they were adequately formalized). It was primarily created to solve mathematical geometry problems and also to play chess. Another example is *ELIZA*, a *chatbo*t, i.e., a program capable of interacting verbally with a person, written in 1966 by Joseph Weizenbaum. ELIZA is a lexical analyzer, and a set of rules (or scripts) capable of simulating a conversation in several languages (English, Welsh, or German). Both G.P.S. and ELIZA provided the first evidence of the possibility of building programs capable of interpreting written and spoken language. (Interestingly, the name ELIZA was inspired by Eliza Doolittle, the protagonist of George Bernard Shaw's *Pygmalion*.) These successes, along with support from leading researchers, convinced government agencies such as the *Defense Advanced Research Projects Agency* (*DARPA*) to fund AI research at various institutions. The government was particularly interested in a machine that could transcribe and translate spoken language, as well as process large amounts of data. Optimism was high, and expectations were even higher, to the point that Marvin Minsky (one of the four co-organizers of the Dartmouth College Conference) wrote in 1970 in *Life Magazine*:

> *In three to eight years, we will have a machine with the general intelligence of an average human being.*

The prophecy did not come true, as many others that have become famous in the field of computers. For example, see (https://www.pcworld.com/article/532605/worst_tech_predictions.html). For the general public, it is worth remembering that Minsky was one of the main advisors to Stanley Kubrick for the creation of the famous movie *2001: A Space Odyssey*. Minsky also won the prestigious Turing Award in 1969 for his influence in the birth of AI. McCarthy won it in 1971.

However, while the *proof of concept* was there, there was still a long way to go before the final objectives of natural language processing, abstract thinking and self-recognition could be achieved. In the 80s, AI was relaunched thanks to two factors: an evolution of algorithmic techniques and an increase in funding. John Hopfield and David Rumelhart popularized *deep learning* techniques that allowed computers to learn from experience. Hopfield was awarded the Nobel in Physics in 2024, thanks to these pioneering studies. We will encounter deep artificial neural networks, the basis of deep learning, later. On the other hand, Edward Feigenbaum introduced *expert systems*, which mimicked the decision-making process of a human expert. These programs asked an expert in a certain field how to respond in specific situations and, once they had learned the answers for almost every situation, they could give advice to non-experts. These systems were widely used in industry. The Japanese government heavily funded expert systems and other AI-related projects as part of their Fifth Generation Computer Project (FGCP). From 1982 to 1990, 400 million dollars were invested with the aim of revolutionizing computer processing, implementing logic programming, and improving artificial intelligence. Unfortunately, most of the more ambitious goals were not achieved. However, it could be argued that the indirect effects of the FGCP inspired a new generation of talented engineers and scientists. Despite this, FGCP funding ceased, and AI seemed to fade from the scene.

Paradoxically, in the absence of government funding and without public attention, AI began to thrive once again. During the 1990s and 2000s, many of the fundamental goals of artificial intelligence were achieved. In 1997, world chess champion and grandmaster Garry Kasparov was defeated by IBM's *Deep Blue*, a computer chess program. This highly publicized match marked a significant milestone in the development of AI-powered decision-making programs. That same year, Dragon Systems' speech recognition software was integrated into Windows, representing another breakthrough—this time in the interpretation of spoken language. It seemed as though there was no challenge that machines could not tackle. Even human emotions appeared to be within their reach, as demonstrated by *Kismet*, a robotic head created in the 1990s at the Massachusetts Institute of Technology (MIT) by Cynthia Breazeal as an experiment in *affective computing*—a machine capable of recognizing and simulating emotions. The name *Kismet* comes from a Turkish word meaning *fate* or sometimes *luck* (https://robotsguide.com/robots/kismet).

A Paradigm Shift

What had changed to allow these spectacular advances? A key explanation lies in the improvement of technology. The fundamental limitation of computer memory that had blocked us 30 years earlier was no longer a problem. Moore's Law, a conjecture formulated in 1965 by Intel co-founder Gordon Moore, which stated that the memory and speed of computers double every 18 months, had finally reached and, in many cases, surpassed our needs. This is exactly how Deep Blue was able to defeat Gary Kasparov in 1997, just as AlphaGo was able to defeat the Chinese Go champion, Ke Jie, in 2017. This created, for some, the illusion that the roller-coaster effect characterizing AI research's evolution from the beginning could be explained: we exhaust the capabilities of AI at the level of our current computing power (memory and processing speed), and then wait for Moore's Law to catch up with us again. But, above all, this progress was essentially due to a paradigm shift in the approach to AI. While in fields requiring precise characterizations, such as chess, algebraic manipulation, and business process automation, artificial intelligence had made significant progress, in other areas like language translation and object visual recognition, the intrinsic ambiguity hindered development. These formal and rigid systems succeeded in contexts where tasks could be performed by encoding clear rules. From the late 1980s through the 1990s, the field of AI entered a phase known as the "AI winter." When applied to more dynamic tasks, artificial intelligence had proven fragile, producing results that, by far, did not pass the Turing Test, meaning they did not reach or imitate human performance. In the 1990s, the turning point occurred. Researchers realized that a new approach was needed, one that would allow machines to learn autonomously, leading to a paradigm shift: moving from the idea of encoding human intuitions into machines to delegating the learning process itself to the machines. Although machine learning had existed since the 1950s, new discoveries enabled practical applications. The most effective methods in this field extract patterns from large datasets through neural networks. For example, to recognize the image of a cat, researchers understood that a machine must learn different visual representations of cats by observing animals in various contexts. In this way, what matters for machine learning is the plurality of different representations of an object, rather than its ideal (or, worse yet, its definition). Modern AI algorithms have the ability to learn types and patterns, such as clusters of words that often appear together when analyzing natural language, or features more frequently found in the image of a cat, and then make sense of reality by identifying networks of similarities and analogies with what the AI already knew. While AI will likely never know in the same way a human mind does, an accumulation of correspondences with patterns in reality could approximate and sometimes surpass human perception and reasoning performance.

This is how the modern field of machine learning (ML) took shape, consisting of programs that learn through experience. Parallel to the development of the algorithmic strategy that led to the modern approach to machine learning, scientists continued to reflect on how ambitious one could be in pursuing the dream of artificial intelligence. Along this line of thought, AI has been defined in many ways, depending on the goals set to be achieved. The definitions that have emerged over time can be summarized in the following terms.

Narrow AI, also called *weak AI*, refers to the automation of specific tasks, but it does not have the ability to develop understanding beyond the scope for which it was programmed. Examples include recognizing an animal in a photograph or translating a text into another language. Other examples include voice assistants like Siri and Alexa, recommendation systems, and self-driving cars.

Strong AI, also known as *general AI*, on the other hand, implies the ability to learn and think *like* humans; that is, it refers to systems capable of understanding, learning, and applying knowledge across a wide range of tasks, with performance similar to human cognitive abilities. Currently, weak AI can, in several fields, even surpass humans in the specific tasks it was designed for, but it operates with many more constraints than even basic human intelligence. All available AI today can be considered weak AI. We can therefore state that strong AI does not yet exist (Togelius 2024).

Finally, *Artificial Superintelligence* is a completely theoretical concept in which AI would surpass human intelligence in all aspects, including creativity, solving any class of problems, and would possess social intelligence. It represents a point at which machines not only match but vastly exceed human intellectual capabilities.

General artificial intelligence is an ambiguous and misleading concept, and so is the concept of superintelligence, especially since we don't have a clear definition of "intelligence" or "generality." Although we continue to make impressive progress in AI technology, we are still far from having a single system capable of performing all the tasks we associate with the idea of intelligence.

Other examples of proposed development environments within AI include *Reactive Machines* and the so-called *Theory of Mind*. Reactive Machines are basic AI systems that react to specific inputs with pre-programmed responses. They do not have memory-based functionality or learning capabilities. An example is IBM's Deep Blue, programmed to play chess and defeat champion

Garry Kasparov. The Theory of Mind is a type of AI still in research, with the ambition to understand and replicate human emotions and thoughts. This type of AI would be able to perceive, interpret, and respond to human emotions and social interactions, allowing for more nuanced and adaptive interactions with humans.

These classifications outline the development and objectives of AI research, charting a path from specific and simple systems to complex and self-aware entities. The shift from weak to strong AI would entail significant advances in understanding human cognition, machine learning, and ethical considerations.

When reading and writing about AI, the context in which the term is used is relevant, and it is important to remember that the very concept of AI is still the subject of broad debate among engineers, scientists, and philosophers. We will give it more concrete form in the two chapters that follow. The future of AI will offer space for multiple perspectives and different technical approaches.

2

Shining Light on AI

It is interesting to observe how widespread confusion surrounds the perception of what AI is—or what it is not. Many processes that elude our understanding are often labeled as artificial intelligence, and at times, any action involving an algorithm (another greatly misunderstood concept of our time) is mistakenly considered AI. Even more misleading is the frequent tendency to equate AI solely with ChatGPT, a generative artificial intelligence system, while in reality, this represents only a subset of the vast AI landscape. This confusion is somewhat understandable, at least in part: as previously noted, the very definition of AI has evolved multiple times over the years. However, given the relevance of this topic in today's society, it is important to establish some clarity.

Algorithms That Learn on Their Own

One key element that I consider distinctive and essential to AI today is *the presence of machine learning, that is the capability of algorithms to learn autonomously*. Traditional algorithms are sequences of instructions, typically formulated using mathematical entities and operations, that follow predefined rules. They operate deterministically, meaning their behavior and output are predictable and repeatable, governed strictly by input data and instructions. These algorithms do not change or adapt based on past experiences or new data.

In contrast, AI algorithms based on machine learning modify their behavior and outcomes according to the data they receive. This means they possess an ability to "evolve" or "adapt," which traditional algorithms lack. AI employs techniques such as supervised, unsupervised, and reinforcement learning to improve over time (which we will explore in the next chapters). It can handle complex data, make predictions, and identify patterns without needing explicit instructions from a human programmer for every action. AI systems learn and adapt. They generalize from pre-existing data and models to apply their knowledge to new cases or scenarios they have never encountered before. This *adaptive learning* is crucial for tasks such as, e.g., image recognition, natural language processing, and personalized recommendations.

For the past 50 years, machines have been unable to demonstrate intelligence according to the Turing Test, but this limitation now seems close to being overcome. Until just a few decades ago, computers functioned strictly based on predefined codes, leading to rigid and static outcomes. Traditional programs could handle vast amounts of data and perform complex calculations, but they struggled with recognizing simple objects in images or adapting to imprecise input. The uncertainty and conceptual nature of human thought posed a significant challenge to the development of AI. However, in the past two decades, technological advances have given rise to AI systems that are beginning to match, and in some cases surpass, human performance in these areas. For example, generative AI algorithms do not simply translate texts by substituting words; they identify and use idiomatic expressions and linguistic patterns. Furthermore, these algorithms are dynamic, evolving in response to ever-changing situations, and can discover new solutions that even humans may not anticipate.

Unlike many traditional ones, modern AI algorithms operate in a nonlinear and, in some cases, stochastic manner (i.e., relying on probabilistic processes), which enables them to identify complex and non-obvious relationships in data. Similar to a classical algorithm, a machine learning algorithm consists of a series of precise steps. However, unlike classical algorithms, these steps do not lead directly to a specific result. They deviate from the precision and predictability of traditional algorithms. This flexibility is even more evident in Deep Learning, where deep neural networks (which we will encounter shortly) "learn" highly complex patterns that would be impossible to define a priori through rigid rules. These characteristics reflect the adaptive nature and autonomy of modern AI algorithms in learning from data.

If we accept that the defining feature of AI today is its ability to autonomously learn, then we can draw the following distinctions.

Navigation systems, such as GPS-based car navigation devices, are not necessarily AI. When we enter a destination, the navigator determines the route by solving a complex combinatorial optimization problem behind the scenes (say, in the cloud). Among all possible road combinations, it selects the one that optimizes the chosen criterion—minimizing travel time or distance, for example. However, this process does not involve a learning phase that allows the algorithm to adapt based on previous choices or the behavior of other drivers. Although these systems use vast amounts of data, they do not learn autonomously because they satisfy predefined criteria—such as, e.g., the Dijkstra or A* algorithms (Hart et al. 1968; Dijkstra 1959).

Similarly, while advanced, many *driver-assistance systems*, such as adaptive cruise control or parking sensors, are not true AI systems. They rely on predefined rules and sensors rather than actual learning from data. Their algorithms calculate trajectories, braking times, and angles, often based on deterministic motion laws, but they do not learn autonomously. This differs from self-driving car algorithms, which do incorporate machine learning.

Even *autopilot systems* in airplanes, which assist pilots in all flight phases except takeoff and landing, are not (necessarily) AI. The same applies to real-time video or image sharing on our mobile phones. These processes rely on image compression and decompression algorithms, such as JPEG or MPEG, which are based on fixed mathematical rules rather than learning from data.

On the other hand, email *spam filters* qualify as AI because they continuously learn based on our past choices regarding which messages we consider spam or unwanted.

Similarly, *customer purchase prediction* algorithms fall under AI, as they use machine learning models to analyze past behavior and forecast consumer preferences.

Churn analysis algorithms are another example of AI. These predictive models identify customers—such as those in telecom or energy sectors—who are at risk of leaving, using historical data that updates as new information is collected.

Other AI-driven technologies include *speech recognition algorithms* in virtual assistants (e.g., Siri, Alexa) that improve their accuracy over time by learning from user interactions; *facial recognition systems* that enhance their precision by continuously training on new images, and *recommendation engines*, such as those used by Netflix or Amazon, which learn from user behavior to offer increasingly personalized suggestions.

The notion that AI is simply any process governed by algorithms is fundamentally incorrect. In particular, algorithms derived from mathematical models based on physical laws (which we will explore later) generally do not involve machine learning, and thus do not qualify as AI. These models are not trained on data but instead implement mathematical equations that encode fundamental laws describing natural, biological, economic, and social processes. Some notable examples include weather forecasting models, aerodynamic and structural analysis models (e.g., crash analysis in the automotive industry), and seismic impact simulation models for high-risk earthquake zones.

While AI continues to evolve, distinguishing true learning-based AI from conventional rule-based algorithms remains essential for an accurate understanding of this rapidly advancing field.

Success Stories

Artificial vision, speech and text recognition, generative AI (for text, images, and films), robotics, autonomous driving for vehicles and public transportation, and expert systems—those capable of solving problems with the expertise of a human specialist—are just some of the domains where AI has achieved remarkable success. See Fig. 2.1 for reference. Let's briefly explore these fields.

Artificial Vision is a branch of technology that enables machines to see, interpret, and respond to their visual environment. Cameras and sensors capture images, which are then processed using algorithms. By combining optics

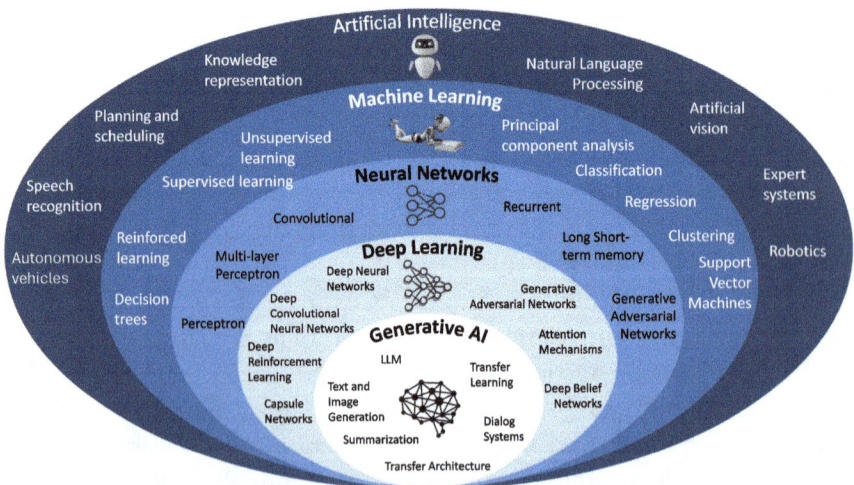

Fig. 2.1 Artificial intelligence, machine learning, neural networks, deep learning, generative intelligence. (Individual images from Shutterstock.com)

and computing, artificial vision systems can perform complex visual tasks that once required human intervention. This process consists of several stages. First, using cameras and sensors, artificial vision systems capture images of their surroundings, ranging from simple photographs to complex 3D representations. These images are then analyzed by algorithms that detect patterns, recognize objects, measure distances, and classify elements. The final stage involves interpreting the analyzed data and making decisions based on the extracted information. In an industrial setting, for example, this could involve sorting products or detecting defects in manufacturing. The quality of the captured images depends significantly on the optical components and sensors used, which determine the precision and efficiency of artificial vision systems. Moreover, the software responsible for processing and analyzing these images plays a crucial role, ensuring that the system can handle vast amounts of data in real time. In some cases, the term *neural vision* is used to highlight the integration of AI in artificial vision, enhancing its capabilities through machine learning techniques.

Robotics is a multidisciplinary field that integrates engineering, computing, and technology to design, build, and operate robots. These machines are typically designed to replicate or enhance human activities, performing repetitive, dangerous, or highly precise tasks. The fundamental components of robotics include mechanical design, which defines the robot's structure and movement capabilities, often inspired by human or animal forms; electrical systems, which include power sources and control mechanisms that enable functionality; and software and algorithms, which allow robots to process information, make decisions, and execute tasks. Although there are many types of robots, they can generally be categorized into: industrial robots, used in manufacturing and assembly lines for tasks like welding, painting, and component assembly with high precision; service robots, designed to assist humans in various fields, including medical robots for surgery and rehabilitation and household robots for cleaning and maintenance; autonomous robots, capable of operating without human intervention, such as drones and self-driving vehicles; humanoid robots, designed to mimic human appearance and behavior, often used for research, entertainment, or companionship. Robotics is an ever-evolving field, driven by technological advancements and expanding applications across multiple sectors. It has the potential to revolutionize the way tasks are performed and enhance both human and machine capabilities.

Expert Systems are computer programs designed to solve complex problems and provide decision-making capabilities comparable to those of a human expert. They rely on both factual knowledge and heuristic reasoning, mimicking the cognitive processes of a specialized professional. These systems derive their name from their ability to store and apply expert knowledge within a specific domain, enabling them to tackle complex issues within that field. An expert system typically consists of three key components: user interface—it facilitates interaction between the user and the system; inference engine—it processes information and applies logical reasoning; knowledge base—it stores domain-specific knowledge, collected from human experts.

The first expert system, developed in the 1970s, was one of the earliest successful applications of AI. These systems are often designed for specialized fields such as medicine, where they can assist in diagnostic processes. Their performance depends on the breadth and quality of the stored knowledge—the larger the knowledge base, the more accurate and effective the system becomes. One common example of an expert system is Google's spell-check and autocomplete suggestions. It is important to note that expert systems are not designed to replace human experts, but rather to support decision-making in complex scenarios. However, they lack human-like reasoning and can only operate within the constraints of their predefined knowledge base.

Speech Recognition (also known as automatic speech recognition or speech-to-text conversion) is a technology that allows computers to convert spoken language into written text. It is often confused with voice recognition, but the two serve different purposes: speech recognition focuses on transcribing spoken words into text; voice recognition is used to identify a specific speaker's voice, often for security purposes. Numerous speech recognition applications and devices are now available on the market. These systems integrate elements such as grammar, syntax, and audio signal processing to understand and interpret human speech. The ideal speech recognition system continuously learns and improves through repeated interactions. Advanced systems also enable businesses to customize the technology to meet specific needs, including language adaptation, speech nuances, and brand recognition. Some key advancements in speech recognition include: linguistic weighting, for prioritizing frequently used words; speaker labeling, for identifying individual speakers in multi-person conversations; acoustic training, for adapting to background noise (e.g., in a call center) and adjusting to various speech styles (e.g., pitch, volume, and speed).

Natural Language Processing (NLP) is an interdisciplinary field focused on enabling computers to process human language data. It is closely related to information retrieval, knowledge representation, and computational linguistics. Typically, natural language processing involves processing large corpora of text using rule-based, statistical, or neural network-based approaches for machine learning. NLP has played a foundational role in generative AI, powering search engines, customer service chatbots, voice-operated GPS systems, and digital assistants. It has also become increasingly valuable in business solutions, helping automate operations and enhance productivity. Until the 1980s, most natural language processing systems relied on manually written rule sets. However, by the late 1980s, the field underwent a revolution with the introduction of machine learning-based algorithms. This shift was made possible by the increasing computational power (aligned with Moore's Law) and the decline of Chomskyan linguistic theories, which had traditionally discouraged corpus-based linguistic analysis (Chomsky 1986). One of NLP's most successful applications is machine translation between languages. This task requires a deep understanding of linguistic structures, contextual meaning, and cultural nuances, making it a challenging and fascinating field within NLP. Instead of relying solely on carefully translated texts, NLP models use parallel corpora—collections of thematically similar texts in different languages. This method, akin to immersion learning, prioritizes exposing the system to large amounts of data rather than focusing solely on precise translations. The success of parallel corpus training has significantly improved machine translation systems like Google Translate, which achieved a 50% improvement in translation quality compared to earlier methods.

Generative Intelligence is by far the area that has garnered the most interest in recent years. While text translation and image classification involve interpreting existing content, generating new text, images, and sounds represents a different challenge. Generative neural networks can create original content. For instance, while a standard neural network can recognize a human face in an image, a generative neural network can create an entirely new face that appears real. A common training method for generative AI involves Generative Adversarial Networks (GANs), where two networks compete in a complementary learning process: the generator creates possible outputs; the discriminator evaluates these outputs and filters out low-quality ones. During training, the generator and discriminator are alternately optimized, each improving the other. Trained on vast datasets—mainly sourced from the Internet—generative neural networks can convert text into images and vice versa, expand or condense descriptions, and perform other similar tasks.

At times, these models yield results that seem surprisingly "intelligent", while at other times, their outputs may appear nonsensical or incoherent. Due to their transformative potential across various fields, including the creative industries, generative AI is attracting significant interest as researchers and developers explore its strengths, limitations, and potential applications.

3

Machine Learning: When Computers Learn (On Their Own!)

As we have seen, the roots of AI lie in a fusion of ideas and visionary efforts, driven by pioneering reflections on the possibility of endowing machines with intelligence. From these origins, AI has undergone a rapid developmental trajectory, constantly redefining the boundaries of our interaction with intelligent technology. However, despite its 70-year history, the spectacular progress we have witnessed over the last quarter-century is also due to two other essential factors: the extraordinary power of computers, particularly those based on GPUs (graphics processing units), which are especially efficient in training and implementing machine learning algorithms, and the vast availability of data—Big Data—which, in many fields, is essential for training machine learning algorithms (see Fig. 3.1).

When we talk about machine learning—the ability of computers to learn autonomously—the most common perception goes something like this: we feed them large amounts of training data so they can learn to perform tasks without being explicitly told how to do them. In other words, without writing specific code for each task. Almost like training puppies. But is that really how it works?

Machine learning is a branch of artificial intelligence that explores how to computationally simulate (or even surpass) human intelligence. It drives most of today's AI advancements by focusing on a single goal: using algorithms to automatically improve the performance of other algorithms.

Here's how it works in practice, using supervised learning, one of the most common forms of machine learning. The process begins with a specific task—for example, "recognizing which photos in a given dataset contain a cat." The goal is to find a mathematical function, known as a *model*, that can

DATA ⟶ **ALGORITHMS** ⟶ **COMPUTERS**

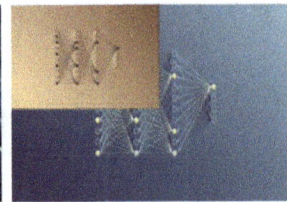

- IoT sensors
- Corporate databases
- Cameras
- Internet

- Deep Neural Networks
- Convolutional Neural Networks
- Large Language Models
- Automatic differentiation
- Backpropagation

- Supercomputers
- Cloud computers
- GPU

Fig. 3.1 The three key factors behind the recent explosion of AI. (Images from Shutterstock.com)

accomplish this task. This model takes numerical inputs—in this case, digitized photographs—and transforms them into numerical outputs that represent labels such as "cat" or "not a cat", which, for simplicity, we can denote as 1 or 0.

The model is built on a mathematical structure defined by a set of numbers called *parameters*, but at the beginning of the process, it is unlikely to produce accurate results.

How AI Models Are Trained

Now it's time to *train* the model, and this is where another type of algorithm comes into play. First, a different mathematical function (called an *objective function*, also known as a *cost function* or *loss function*) calculates a number that represents the current "distance" between the model's output and the desired result for all the images in the initial dataset, which we will call *the training set*. In this specific example, this number indicates the percentage of times the model fails to correctly identify photos of cats. Next, the training algorithm uses the objective function's distance measurement to adjust the original model (i.e., the parameters that characterize it). It does not need to "know" what the model represents—in other words, it has no awareness of what it is doing. It simply updates the values of the parameters so that the objective function (and therefore the model's percentage error) becomes as small as possible.

3 Machine Learning: When Computers Learn (On Their Own!)

Once these adjustments are made, the process repeats, leading to successive iterations. The updated model transforms the training set inputs into (slightly improved) outputs, then the objective function suggests another (slightly improved) adjustment to the model. After a sufficient number of iterations, the trained model should be able to produce accurate outputs for most of its training examples. The ultimate goal, however, is to maintain that performance on new examples of the task, as long as they are not too different from those in the training set.

This automated process brings forth a mathematical representation of the task without requiring humans to specify which details are important. With an efficient algorithm, a well-chosen objective function, and a sufficiently rich training set, machine learning can generate powerful computational models that perform tasks very effectively. A mathematical description of the entire process will be outlined (without too many technicalities!) in the next chapter.

Classification and prediction tasks—such as identifying cats in photos or detecting spam in emails—are usually based on *supervised learning*. This means that the training data is already labeled in advance, and the label corresponds to the desired output: for example, photos containing cats are labeled as "cat" or, more conveniently, assigned the number 1, while all other images are labeled as 0. The training process then determines the mathematical function (*the model*) that can map as many inputs as possible to their corresponding (known) outputs. Once trained, the model will be able to label unknown examples without making too many errors. For instance, in early 2020, researchers at MIT used supervised learning to discover a new antibiotic, halicin. To achieve this, they employed a database containing 2000 molecules to train a model, where the molecular structure was provided as input and the antibiotic's effectiveness as output (Marchant 2020).

Unsupervised learning, on the other hand, identifies structures within unlabeled examples, grouping them into clusters that are not predefined. In this case, the input data has no labels. It is often used for raw datasets and is responsible for converting them into structured data. In many contexts, large amounts of data are generated, including log files produced by computers. For this reason, unsupervised learning represents one of the biggest challenges in machine learning. Content recommendation systems that learn from a consumer's past behavior, as well as certain object recognition tasks in computer vision, can rely on unsupervised learning. For example, video streaming services like Netflix use algorithms to identify groups of users with similar viewing habits, allowing them to suggest additional content to watch.

Some tasks, such as language modeling performed by large language models like GPT-4—which we will encounter later—use intelligent combinations of supervised and unsupervised techniques, known as self-supervised and semi-supervised learning. *Semi-supervised learning* occurs when the training dataset contains both labeled and unlabeled elements. Consider a neurosurgery department that acquires a vast number of tomographic images, but due to time constraints or lack of expertise, only a limited number have been annotated by a neurosurgeon with a description of the observed pathology. This means specifying whether or not a tumor is present and, if so, identifying its type.

Finally, *reinforcement learning* generates a model function using a reward signal instead of examples of desired outcomes. Its goal is to teach *agents*—entities capable of making decisions and taking actions—to adopt a sequence of decisions within an environment to optimize total cumulative rewards. By maximizing this reward through trial and error (carried out thousands, tens of thousands, or even more times at incredible speed thanks to powerful computers), a model can improve its performance on dynamic and sequential tasks, such as playing checkers, chess, or Go, as well as controlling the behavior of real and virtual agents (such as self-driving cars or chatbots). The primary objective of reinforcement learning is to enable an agent to interact with its environment, observe the consequences of its actions, and adjust its behavior accordingly. However, reinforcement learning requires human involvement in creating the AI's training environment: humans define a simulator and a reward function, and the AI trains based on these elements. To achieve meaningful results, carefully selecting the simulator and the reward function is crucial.

In 2017, *AlphaZero*, the program developed by Google DeepMind, defeated *Stockfish*, which was then the most powerful chess program in existence. This event marked a significant moment in the history of both chess and artificial intelligence, as AlphaZero demonstrated the ability to learn and develop game strategies without any human supervision. To accomplish this, it used a reinforcement learning program that played against itself. To evaluate its performance, it used a reward function that assessed its moves based on the opportunities they created.

Mathematical systems based on machine learning are often extremely powerful and accurate; however, they are not free from structural weaknesses. The most common issue is *overfitting*: an *overfitted* model builds a function that fits the training data so well that it cannot reliably generalize to represent new, unseen data. For example, imagine a machine learning algorithm trained to determine whether an input image contains a cat. If the model fails when the cat in the image is upside down—because it was only trained on upright cat images—this would indicate overfitting. Another cause of overfitting occurs when the training data contains too much inaccuracy or *noise*, leading the model function to become overly complex in an attempt to capture this excessive variability. In such cases, the model fails to *generalize*, meaning it cannot understand the true nature of data beyond its training set. To illustrate this further, suppose we train a model to predict the average daily temperature over 3 months of spring in a specific location. If the training data is affected by noise in the measurement process, the model might fail to capture a generally increasing temperature trend from March to June, which one would typically expect.

Bias in data is often due to *non-representative sampling*. It can arise from distortions or systematic errors in the way data is collected, selected, represented, or interpreted. For instance, if all the recorded side effects of a drug came from a study conducted exclusively on male subjects, the model might struggle to generalize its predictions to the female population. In other words, biases in training data can be amplified during the learning process, leading to distorted—or even incorrect—results. Moreover, even when a model performs well, it is not always clear *how* it works and *why* it works. Deep learning algorithms, which we will discuss later, are particularly affected by this problem of *interpretability and explainability*.

Back to Definitions

Let's take a step back and put the definitions and concepts in the right order. Perhaps the oldest definition of machine learning comes from an apocryphal quote attributed to Arthur Samuel, often linked to his 1959 paper on machine learning for checkers:

> *Machine learning is the field of study that gives computers the ability to learn without being explicitly programmed to do so.*

A more concise definition was proposed in a 2017 report by the Royal Society:

> *Machine learning is a set of rules that allows systems to learn directly from examples, data, and experience.*

In a review article by M. Jordan and T. Mitchell (2015), the term *learning* is defined as the process of transforming information into skills or knowledge, while *machine learning* is described as *automated* learning. The machine learning process can be understood as an algorithm that takes information as input and produces knowledge as output. As previously mentioned, machine learning algorithms generally fall into four main categories: supervised learning, semi-supervised learning, unsupervised learning, and reinforcement learning.

Machine learning is particularly effective when large amounts of data are available and when developers have access to the ground truth, meaning they can verify the correctness of the results, often even during model training. However, its effectiveness decreases when datasets are scarce, incomplete, or unlabeled, which is often the case when data acquisition is costly. It also struggles when the ground truth is unknown and no reference datasets exist, or in situations requiring decisions where interpretability is crucial, meaning the model's outputs must be understandable by humans for high-impact decision-making.

In the case of supervised machine learning algorithms, their main tasks can be identified among the following:

Regression: This involves extracting a function from a discrete distribution of data (inputs and outputs) that describes its behavior. It is the so-called data-based mathematical modeling. For example, identifying a function that describes how the value of a certain stock changes over time on the Stock Exchange. In this case, the training set consists of N pairs of input data and corresponding ground truth values, that is, N different dates and the corresponding prices of the specific stock. Another example could be representing the temperature behavior at a certain location with a curve, starting from a daily temperature distribution.

Classification: This involves predicting a value from a finite set, for example, the label to associate with an image. The training set consists of pairs of input data and ground truth values, in this case, a label chosen from a finite set. The standard approach involves assigning a score to each potential class, with the correct class receiving the highest score. This is applied, for example, to image classification problems (cat or not cat), spam filtering (email to delete or keep), sentiment analysis (aimed at determining if the content of a text expresses a positive, negative, or neutral sentiment), etc.

In the case of unsupervised machine learning algorithms, we can highlight the following tasks:

Clustering: This refers to grouping data so that similar data points are in the same group, even without predefined labels. For example, with a certain distribution of points in a Cartesian plane, the goal is to find subsets that share specific similarities. If these were individuals, they could be grouped based on pre-assigned income brackets. This is used for community detection in biology, social sciences, economics, customer segmentation for specific purchases, etc.

Associations: This refers to identifying relationships between variables in the data, such as in market analysis, where correlations are sought. This approach is used to discover the probability of co-occurrence of elements in a data collection. A classic example is recommendation systems that analyze which products are frequently purchased together, as seen in market analysis.

Density Estimation (e.g., anomaly detection, generation of realistic scenarios): This task focuses on finding common patterns within a population of data without assuming a rigid structure. It is useful for detecting anomalies (data that does not conform to expected patterns) or generating realistic scenarios similar to those described by the training data.

Dimensionality Reduction: This task involves finding a compact representation of complex data by reducing its dimensionality while retaining as much of the relevant information as possible. It is very useful for addressing high-dimensional data problems, making them easier to visualize and manage.

The versatility across multiple tasks makes machine learning algorithms especially appealing and interesting.

4

Mathematics...the Bare Minimum

Before we venture into what will likely be the most challenging section of this book for many readers, I think it's important to clarify a crucial point. By now, after so many introductions, it should be clear that machine learning algorithms are a powerful machine built on as much data as possible (preferably Big Data). Data, are "given", meaning they are known quantities! And it's essential not to confuse them with solutions! The solutions, on the other hand, are unknowns, and they must be found. Every time we face a mathematical problem, the challenge is to go from data to solutions, i.e., from input to output. How to do that is what we are discussing in this book. We can think of machine learning as a process that allows us to associate a certain set of data with the corresponding solution. Whether it's assigning a label that indicates whether a particular photo shows a dog or something else, or recognizing whether a certain medical image (such as a CT scan or MRI) hides the presence of a tumor, and if so, which type of tumor it is, or partitioning a set of individuals into subsets with similar characteristics (such as their spending capacity or tendency to develop specific diseases), or even forecasting the weather for the upcoming weekend based on current data like wind speed, atmospheric pressure, and humidity. As can be guessed from these examples, the types of data can vary greatly, and the solutions that can be associated with them will also be very different. What remains unchanged is the mathematical structure: the trinomial

$$data \rightarrow mathematical\ procedure \rightarrow solutions.$$

© The Author(s), under exclusive license to Springer Nature Switzerland AG 2025
A. Quarteroni, *Artificial Intelligence*, Copernicus Books,
https://doi.org/10.1007/978-3-031-92973-1_4

The mathematical procedure, in turn, can take many different forms. In the case of partitioning subsets as mentioned earlier, statistical clustering methods can be used, while in meteorology, a complex system of equations that mathematically translate the physical processes of our atmosphere must be used.

The Building Blocks of ML Construction

Machine learning can also be considered a particularly notable example of a mathematical procedure. To understand how the machine learning process unfolds, we can refer to the general formulation proposed by Tom M. Mitchell back in 1997:

> *A program learns from an experience E with respect to a specific goal T (task) and a given performance measure P if its performance with respect to T improves with experience E.*

As can be seen, the formalization of the process starts to take shape, although in a rather abstract form. To make this statement more concrete, let's reconsider the now well-known image recognition problem: we want to write a program that, for a given image called x, should return a binary number y which will be 1 if the starting image is of a cat and 0 otherwise. We can imagine that a machine learning algorithm works roughly in this way:

(E): We start with a *training set* consisting of N images x_i representing cats or something else (other animals, objects, etc.) and a corresponding value y_i for i ranging from 1 to N.

(M): We introduce a possible *model*, that is, a function that associates a generic image x with a binary value y (either 0 or 1). In symbols, $y = f(x, p)$, where p represents values to be determined (generically referred to as *parameters*), and f is the function that associates the pair (x, p) with the result y. How f depends on x and p identifies the chosen model.

(P): We *train* the model by selecting from all possible parameters the ones that minimize the error when the model is applied to the training set images. In other words, the values $t_i = f(x_i, p)$ should differ as little as possible from the true values y_i, for all values of the index i ranging from 1 to N. We therefore compute the optimal parameter

$$p_{opt} : J(p_{opt}) = \min_{p \in \mathbb{R}^n} J(p)$$

One possible criterium is to choose *J(p)* as the sum of the squares of the discrepancies between t_i and y_i, that is to set

$$J(p) = \sum_{i}\left[y_i - f(x_i;p)\right]^2$$

However, there are naturally other possible definitions of error. As a matter of fact, the choice of function *J(p)*, called *cost function*, but also *objective function* or *loss function*, in the previous minimization formula characterizes the *performance measure P* of our model.

For those who dislike mathematical formalism, beyond the technicalities, we could say that in this case, with machine learning, we propose a model of the relationship between any image *x* and its semantic value (*y*, which, in this case, indicates whether the image is or isn't of a cat) dependent on parameters that are chosen in such a way as to optimize the performance of the model when applied to the training images.

At this point, once the numerical values of the parameters are determined, and thus the function *f* is fully determined, our model will be ready to be tested on any new image that is not already present in the training set. Naturally, changing the application context, we could now start from any set of inputs *x* (such as people with Covid-19) to generate an output *y* (such as the estimated number of days required for their recovery).

There are various criteria for carrying out the experience phase (E), which, as we've seen, are usually referred to as unsupervised, supervised, semi-supervised, or reinforcement learning. In unsupervised learning, only the x_i values—not the corresponding y_i values—are included in the training set (training is done "blindly"). In supervised learning, both are included. In reinforcement learning, the training set is not fixed in advance, but it changes and is dynamically updated based on results and context. In less mathematical but more imaginative terms, we could say that, in general, supervised learning provides specific knowledge during the training phase that helps to build a true database of information and experiences. This way, when the computer faces a new problem, it doesn't have to do anything except draw from the experiences stored in its system, analyze them, and decide which response to give based on pre-coded experiences. Unsupervised learning, on the other hand, means the information provided is not coded, meaning the computer has the ability to access certain information without having any example of how it should be used, and thus without knowledge of the expected outcomes depending on the chosen decision. It will be the computer itself that must catalog all the information it has, organize it, and learn the result that it leads to. Reinforcement learning is the most complex form. It assumes that the computer has systems and tools capable of improving its learning and, most importantly, understanding the characteristics of its surrounding environment. This type of learning is typical, for example, of autonomous vehicles, which benefit from a complex system of cameras, lidars, and supporting sensors. As can be understood, choosing the right model f is crucial for the development of an effective machine learning algorithm. Naturally, there are many options available in this regard.

The simplest models assume a linear (proportional) dependence of f on both x and p. Others assume a non-linear (polynomial) dependence on x, with polynomial coefficients represented by the parameters to be optimized.

Then there are models based on *random forests*, where the choice of response y is based on a series of dichotomous choices. This model consists of multiple decision trees, each of which provides an answer to a specific question. For example: "I should buy an apartment." From there, a series of questions can be asked to determine an answer, such as "How many rooms should it have?" or "What is the available budget?" And again: "In which area of the city should it be?" These questions constitute the decision nodes of the tree, which serve to partition the data. Each question helps the individual make a final decision, indicated by the leaf node. Observations that meet the criteria will follow the "yes" path, while those that do not will follow the alternative path.

Decision trees try to find the best partitioning of data and are typically trained using a specific algorithm called Classification and Regression Tree (CART). A decision tree is an example of a classification problem, where the class labels are "buy" and "not buy." In general, the more decision trees form a set in the random forest algorithm, the more accurate the results will be, especially if the individual trees are not correlated with each other.

We continue by recalling that *Support Vector Machines* fall under the case of supervised learning. Starting from a training set with labeled elements (cat/non-cat, in the initial example), a model is built to assign new examples to one of the two classes, obtaining a deterministic binary classifier. A support vector machine represents the examples as points in space, mapped in such a way that examples belonging to the two different categories are clearly separated by the largest possible margin. New examples are then mapped into the same space, and the prediction of the category they belong to is made based on the subspace they fall into.

Artificial Neural Networks

However, the models that have taken the lead by far are those based on *artificial neural networks (ANN)*. Artificial neural networks are mathematical algorithms that aim to emulate the behavior of biological neural networks, those that are biological, made up of neurons, axons, and dendrites. A biological neuron is a specialized cell in the nervous system, responsible for transmitting information through electrical and chemical signals. We can imagine it as being composed of several parts (for a graphical representation, see Fig. 4.1).

The soma (or cell body) contains the nucleus and other essential cellular structures. It is responsible for maintaining the cell and integrating signals received from the dendrites, branches that receive signals from other neurons. The axon is a long extension that transmits signals from the soma to other neurons, muscles, or glands. It ends with the axon terminals. The myelin sheath is the covering that wraps around the axon (in many neurons) and increases the speed of signal transmission. It is made up of glial cells, such as oligodendrocytes in the central nervous system and Schwann cells in the peripheral nervous system. The axon terminals (or synapses) indicate the end of the axon, where neurotransmitters are released to communicate with other neurons.

As for its functioning, at rest, a neuron exhibits an electrical potential difference across the cell membrane, with the inside of the cell at a lower potential than the outside (about −70 mV). This is maintained by ion pumps, which are transmembrane channels that open at the cell membrane to

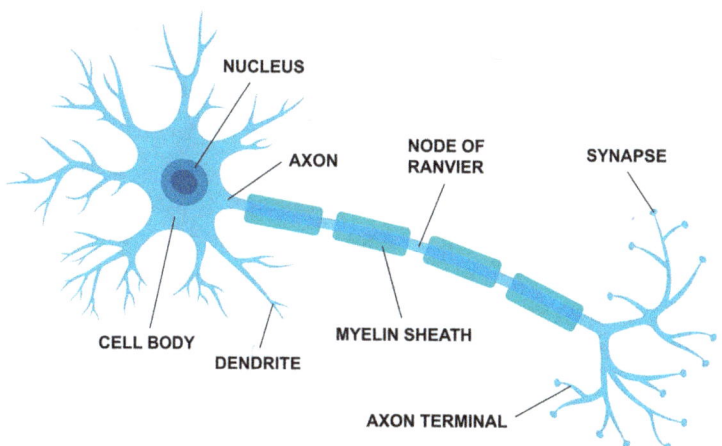

Fig. 4.1 A biological neuron and its components. (Image from Shutterstock.com)

regulate the concentration of sodium (Na+) and potassium (K+) ions. When a neuron receives a signal strong enough to exceed a certain threshold, rapid depolarization of the membrane occurs. Sodium channels open, and Na⁺ ions enter the cell, making the inside more positive. After a brief phase, the sodium channels close, and potassium channels open, allowing K+ ions to exit the cell, restoring the negative potential. The action potential travels along the axon like a wave. In myelinated fibers, the signal "jumps" between the so-called Ranvier nodes (which are areas without myelin) in a process called saltatory conduction, which significantly increases the speed of transmission. At the axon terminal, the action potential triggers the release of neurotransmitters from synaptic vesicles into the synaptic cleft (the space between adjacent neurons). The neurotransmitters bind to receptors on the membrane of the postsynaptic neuron (or other cells like muscles or glands), transmitting the chemical signal. This can lead to the generation of a new action potential in the postsynaptic neuron or other cellular responses.

The neuron is an essential part of communication within the nervous system. Together, neurons form complex networks that enable all brain and bodily functions, from sensory perceptions to motor responses, to complex cognitive processes like thinking and memory.

Artificial neural networks are nonlinear models (compositional, i.e., obtained by applying nonlinear functions in sequence), with their core element being the *perceptron*, the artificial neuron invented in 1943 by Warren McCulloch and Walter Pitts (1943) which simulates the logical behavior of a biological neuron. However, the term perceptron was introduced only in 1958 by Frank Rosenblatt (1958), a researcher at the Cornell National Laboratory. The McCulloch and Pitts model can be seen as a particular case of a perceptron. The goal was to develop a method to encode information similar to how the human brain does, which connects around one hundred billion neurons with quadrillions (10^{21}) of synapses. An artificial neural network encodes relationships between nodes (similar to neurons) and numerical weights that represent the intensity of the connections (via synapses) between the nodes. For decades, the lack of computational power and advanced algorithms slowed the development of neural networks, which remained mostly rudimentary. However, recent advances in both areas have finally freed AI developers from these restrictions.

Similarly to what happens in a biological neuron through the dendrites, an artificial neuron receives input signals from other artificial neurons. Each input is associated with a weight (w) that represents the relative importance of that input. Weights are numerical values that modulate the strength of each input and are generally modified during the learning process. At this point, a function calculates the weighted sum of the inputs, using an expression of the form

$$z = \sum_i w_i x_i + b$$

where x_i are the inputs, w_i are the associated *weights*, and b is a *bias* term that allows shifting the activation function. See Fig. 4.2. The *activation function* applies a nonlinear transformation to the weighted sum to determine the neuron's output. Common activation functions include the sigmoid, ReLU (Rectified Linear Unit), and hyperbolic tangent. Their graphs are shown in Fig. 4.3.

The result of the activation function, which can be used as input for other neurons in subsequent layers of the neural network, constitutes the output of the individual neuron:

$$y = f(x; p) = \sigma\left(\sum_j w_j x_j + b\right)$$

where we have denoted by $p = [w_1, w_2, \ldots, w_n, b]$ the set of parameters that can vary for each individual neuron.

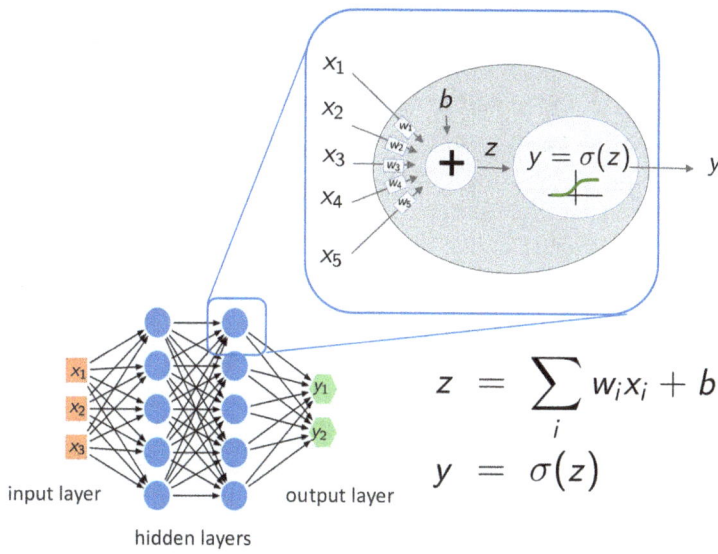

Fig. 4.2 An artificial neuron of a neural network

Fig. 4.3 Examples of activation functions: the Heaviside function (**a**), ReLU (**b**), sigmoid (**c**), and hyperbolic tangent (**d**). (*With the permission of Paola Gervasio*)

In McCulloch and Pitts' original 1943 model, both the inputs and outputs are binary values (0 or 1), the weights were all equal to 1, and the activation function was the Heaviside function (displayed in Fig. 4.3a). We can modify the slope and the position of the transition in the sigmoid function by scaling and translating the argument, or, in neural network terminology, by applying weights and biases to the inputs.

An artificial neural network is a mathematical construction consisting of several layers of artificial neurons. In so-called *Feed Forward* networks, each neuron in a single layer receives input from the previous layer's neurons and provides an output that will feed the neurons in the next layer. The first layer provides the input values x_i, and the final layer provides the output values y_j.

In abstract terms, artificial neural networks represent structures designed to simulate the way our nervous system analyzes and processes information. They are characterized by their self-learning ability, which allows them to produce results that progressively improve as the quantity (and quality) of data available for training increases. The learning process occurs during the training phase, where a set of trainable parameters is calculated so that, for a dataset where both input and output values are known, a sufficiently accurate reconstruction of the underlying law can be obtained. The law that allows the output to be expressed as a function of the input has been previously indicated as $f(x,p)$ where p are the trainable parameters, that is the set of coefficients that characterize the behavior of an artificial neural network, i.e., the matrix containing the weights W and the bias vector b. (A vector is a sequence containing a finite number of data entries—specifically, real numbers—whereas a matrix is an array composed of a finite collection of vectors, all of equal length.) The training process aims to determine these parameters in the most effective way possible. It is essentially a mathematical optimization process, which involves minimizing the *cost function J*, a function that we have introduced before and quantifies the difference between the predicted outputs of the artificial neural network and the actual outputs corresponding to the input data in the training set. The cost function, also known as the objective function or loss function, depends on a large (often extremely large) number of parameters. Minimizing it requires identifying the optimal combination p_{opt} of parameters W and b, where the cost function reaches its lowest possible value. This optimization process takes place in a high-dimensional Euclidean space and relies on sophisticated minimization algorithms, which involve extremely high computational costs. We will revisit this aspect in Chap. 6, where we discuss generative AI and the broader issue of the environmental sustainability of machine learning algorithms.

The number of layers, and the number of neurons per layer, characterize the architecture of the artificial neural network. These are called *hyperparameters*. Other hyperparameters include those that characterize the cost function and the activation functions.

Deep neural networks (DNNs) are composed of a large number of neuron layers. The different layers allow for the representation of complex, interwoven concepts. They can indeed identify complex relationships and intricate connections, including those that may escape human intuition. During the training phase, the weights within the network are adjusted as new data is acquired. As already noticed, as far as the volume of data and the number of layers in the network increase, the parameters W and b begin to reflect the relationships more accurately. Current deep neural networks are composed of about ten layers. Naturally, this number only represents an average order of magnitude, which can actually vary significantly depending on the specific tasks the networks need to perform.

As an example, AlexNet, a convolutional neural network that won the ImageNet competition in 2012, reducing the image classification error from 26% to 15%, used a deep architecture with eight layers and leveraged GPUs (graphics processing units) to accelerate training. Also, the AlphaGo system by DeepMind, which defeated the world champion in Go, an extremely complex game, used deep neural networks alongside other learning techniques to master long-term strategies. By showing a deep learning network an enormous number of food pictures, the network will recognize whether a new picture represents a hot dog or not. Or, if we show it images, videos, and data collected from sensors in a car, it will know how to drive the car autonomously.

Variable Architectures

Naturally, the architecture of artificial neural networks can be highly varied. Generally, it will depend on the specific class of problems to be solved: ANN used to classify an image (cat/man) will be very different from those that need to, for example, calculate the mortality risk factor for patients with electrical heart disorders. ANNs are thus interconnected groups (layers) of nodes (artificial neurons) capable of "learning" to perform tasks such as classification or regression after being trained. There are various application fields in which ANNs have proven effective (often even more effective than human intervention). Just to mention two examples in the medical field, ANNs have already been applied to the classification of arrhythmia from single-lead ECG (electrocardiogram) data from over 50,000 patients, or to the recognition of skin mole danger levels, showing performance comparable to that of experts in both cases.

In *supervised* learning, the main domains of application for deep neural networks include Natural Language Processing (NLP) and Computer Vision. Among the networks for supervised learning, we can mention *Feed Forward Neural Networks* (FFNN), characterized by the absence of cycles in the connections between neurons. They are used for tasks such as classification, regression, and pattern recognition. See Fig. 4.4.

Convolutional Neural Networks (CNN) are designed to work with grid-structured inputs that exhibit strong spatial dependencies in local regions (e.g., 2D images, text, time series, and sequences). They are ideal for feature extraction from inputs, creating similar values from local regions that share similar patterns. Their applications include image classification, object detection, and facial recognition. Examples include: LeNet (1998, used for digit recognition), AlexNet (2012), VGGNet-16 (2014), ResNets (2015), and GoogLeNet (2015). For an example, see Fig. 4.5.

Recurrent Neural Networks (RNN) use loops within the network to maintain a state that captures information from previous inputs, making them suitable for processing temporal sequences. Applications include time series forecasting, language modeling, machine translation, and speech recognition.

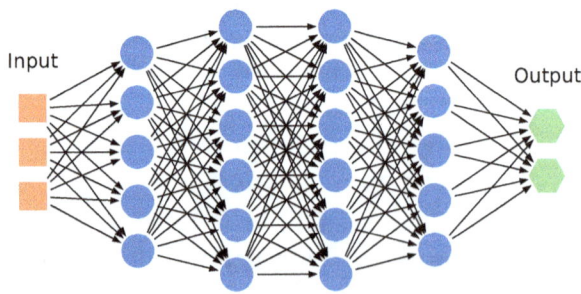

Fig. 4.4 An example of Feed Forward Neural Network

Some examples include: basic RNN (Vanilla RNN) and LSTM (Long Short-Term Memory). A basic example (Vanilla RNN) is shown in Fig. 4.6.

Attention models use so-called self-attention mechanisms to process input data in parallel, rather than sequentially as in RNNs, making them more efficient and scalable. They can focus attention on specific areas of text instead of a window of tokens where all have the same weight, enabling them to "ingest" large amounts of information by capturing important parts. Applications include natural language processing, machine translation, text generation, and image classification. Examples include: Transformer (2017), BERT (2018), GPT (2018), ViT (Vision Transformer, 2020), GPT-3 (2020), PaLM (2022), LaMDA (2022), GPT-4 (2023). BERT, GPT-x, and LaMDA are examples of large language models (LLMs), deep neural networks capable of generating language and performing natural language processing tasks, such as classification.

Reinforcement learning architectures focus on learning decision-making policies through interaction with a given environment. Applications include games such as AlphaGo, robotics, autonomous vehicles, and resource management. Examples include: Deep Q-Networks (DQN, 2015), Policy Gradient Methods, and Actor-Critic Methods.

Now, let's look at *unsupervised* learning architectures.

Generative Adversarial Networks (GANs) consist of two networks, a generator and a discriminator, which are trained together in a zero-sum game: the generator creates fake samples by adding random noise to the training data, while the discriminator tries to distinguish real data from generated data. These networks can generate new content that looks realistic, maintaining statistical similarities. Applications include image generation, image super-resolution, data augmentation, and style transfer, which is the modification of the aesthetic or visual appearance of an image while keeping the structural elements or core content intact. For example, one can transform a realistic

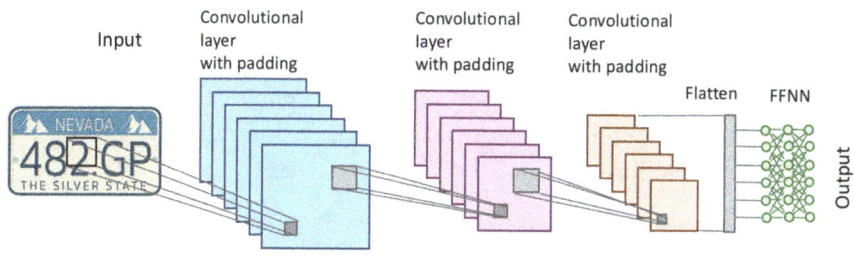

Fig. 4.5 A convolutional neural network for digit recognition. (Nevada plate image from Shutterstock.com)

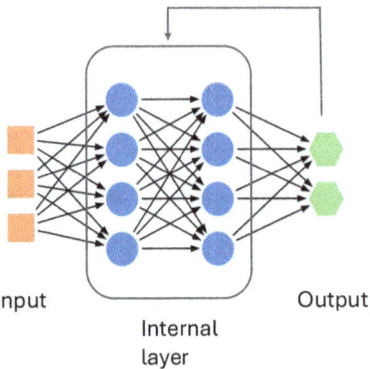

Fig. 4.6 An example of Recurrent Neural Network

photograph into a painting in the style of Picasso or Raphael, or make a modern image look like a comic book illustration. Examples include "Vanilla" GAN, DCGAN, CycleGAN, StyleGAN.

Autoencoders are neural networks trained to reconstruct the original input by compressing it (encoding) and then reconstructing it (decoding). They are used for dimensionality reduction, anomaly detection, data compression, and generative modeling. Examples include Vanilla Autoencoder, Variational Autoencoder, Denoising Autoencoder, and Sparse Autoencoder. A representation is shown in Fig. 4.7.

Graph Neural Networks (GNNs) are designed to work with structured data such as graphs, where nodes represent entities and edges connecting two adjacent nodes represent relationships. An example is illustrated in Fig. 4.8. Their applications include social network analysis, recommendation systems, and protein structure prediction. Notable examples include Graph Convolutional Networks, Graph Attention Networks, and Message Passing Neural Networks.

Finally, *hybrid architectures* (supervised and unsupervised) combine elements from different networks to leverage their strengths. They are used in complex tasks that require a combination of techniques, such as video analysis and multimodal learning, where the training data is heterogeneous, like images, text, and sounds. Examples include: Convolutional Neural Networks (CNNs) combined with attention mechanisms to enhance feature extraction, or hybrid Transformer-CNN networks, used in vision tasks, which combine local feature extraction with global context awareness.

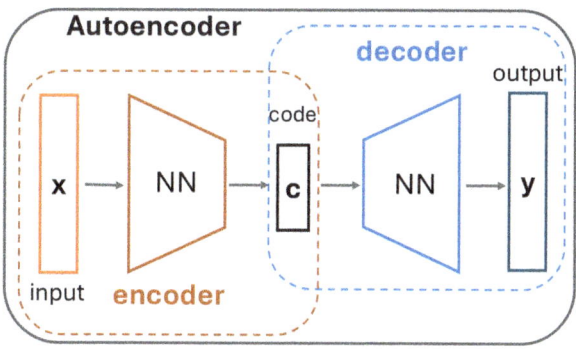

Fig. 4.7 The structure of an Auto-Encoder

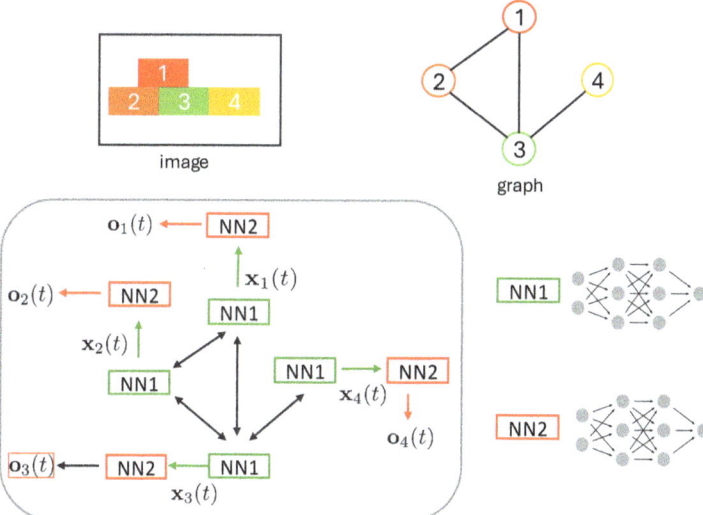

Fig. 4.8 An example of a Graph Neural Network

5

What Data for Learning, and What Kind of Learning?

According to a recent estimate (Reinsel et al. 2017), by 2025 we will be approaching 200 zettabytes of data in the universe's database. One zettabyte equals 10^{21} bytes, or one thousand billion billion bytes. The twenty-first century will therefore be remembered as the era of zettabytes. (Interestingly, 10^{21} is also the estimated number of synapses in the human brain, as we saw in the previous chapter.) If we were to sit comfortably on the couch and watch an HD video containing 200 zettabytes of data, we would need to wait... 40 trillion years before reaching the end of the projection! 200 zettabytes would be enough to fill the memory of 1.5 trillion iPhones or to record 2500 times all the words ever written by every human being. As illustrated in Fig. 5.1, in 1986, when data began to be counted, there were 280 trillion bytes. In less than 40 years, that number has multiplied by nearly a million times! This corresponds to an average growth rate where the volume of data doubles approximately every 18 months. Interestingly, it follows the same dynamics as Moore's Law for the increasing computational power of computers! To further illustrate these production rates, a recent analysis estimates that the global internet population has grown from 2.1 billion to 5.2 billion in just a decade (2013–2023). This surge generates unimaginable amounts of new data every minute of every single day, as shown in Fig. 5.2. The explosive growth of data is driven by several factors. IoT devices alone are expected to generate 90 zettabytes of data per year by 2025. Almost 30% of the global data sphere will consist of real-time data. By the end of 2025, every connected person in the world, approximately 75% of the global population, will interact with data

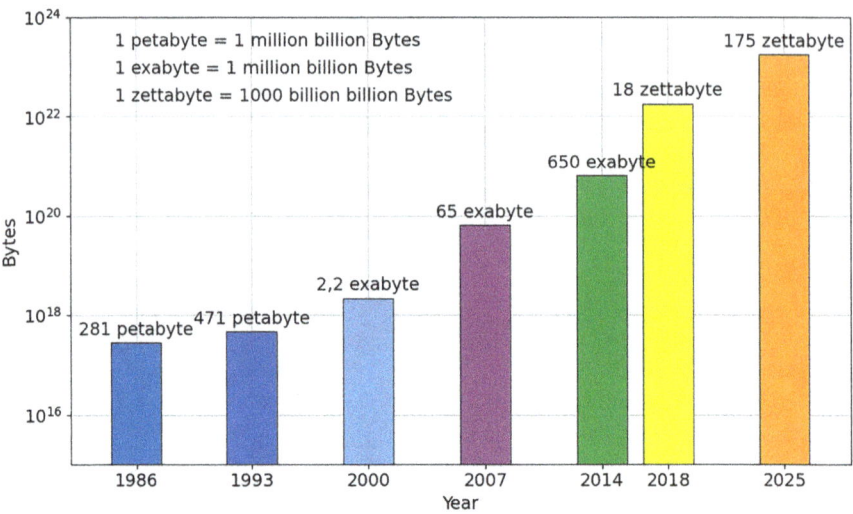

Fig. 5.1 The growth of Big Data over the past 40 years

over 4900 times/day, which means once every 18 s. These trends highlight that the data we generate is not only increasing in volume but also in complexity, frequency, and immediacy. This poses a challenge that requires sophisticated approaches to store, process, and, most importantly, use data to extract actionable insights.

The End of Theory, Anderson's Provocation

This remarkable data explosion is the basis of Chris Anderson's provocative thesis, expressed in his famous article: *"The End of Theory: The Data Deluge Makes the Scientific Method Obsolete"* (Anderson 2008).

Anderson's concept of "the end of theory" generated significant interest, particularly in business and data science academia. At the time, as editor-in-chief of *Wired*, Anderson hypothesized that the abundance of data would render theoretical models, those based on equations derived from physical laws, obsolete. He argued that data correlations are sufficient, making causality irrelevant, meaning there is no need to establish a rigorous cause-effect relationship. Anderson stated that the new availability of vast amounts of data, along with statistical tools to analyze them, offers a completely new way to understand the world. In his view, correlation surpasses causality, and science can progress even without coherent models or unified theories. He claimed that correlation is enough, so we can stop searching for models.

Fig. 5.2 Data Never Sleeps (https://www.domo.com/learn/infographic/data-never-sleeps-11)

Instead of forming hypotheses, we can simply analyze data by feeding numbers into the world's largest computing clusters and letting statistical algorithms find patterns where traditional science cannot. To support his argument, Anderson used the example of Google's search engine, explaining that Google's foundational philosophy is that there is no need to know why one page is better than another. If the statistics of inbound links indicate that it is, then that is enough. There is no need for semantic or causal analysis. Consequently, he concluded that there is no reason to cling to old methods and proposed that it is time to ask what science can learn from Google.

Anderson's assertion that correlation can replace causality contains several fallacies. The idea that Big Data can fully capture any research domain ignores the fact that every dataset is inherently influenced by the platform and ontology, meaning the conceptual representation used. In other words, data reflects a specific viewpoint rather than an absolute truth. Moreover, the assumption that data can exist without human interpretation is misleading, as human decisions shape the collection, analysis, and presentation of data. A striking example of this issue comes from an experiment conducted by the city of Boston, where smartphone accelerometers were used to detect potholes in streets. This initiative, implemented through the Boston Street Bump App, initially appeared to be an innovative and cost-effective solution. However, the data analysis revealed an inherent bias. The app only collected data from drivers with smartphones, which meant that neighborhoods with fewer smartphone owners—often poorer, older communities—generated less data. As a result, wealthier areas appeared to have better-maintained roads, while underprivileged areas seemed to have more potholes, even if that was not necessarily the reality. Despite efforts by Boston's New Urban Mechanics office to address this issue, less conscientious policymakers overlooked the data bias and misallocated resources, further exacerbating social inequalities.

The key takeaway from this example is that while data can bring knowledge, it can also create false knowledge. Proper data analysis requires a strong theoretical foundation, a topic that will be explored further in Chap. 11. There, we will see that some of the greatest scientific discoveries of the past century would have been impossible with data alone.

Data… Matters

Machine Learning systems handle structured data in the form of vectors or matrices, which may evolve over time, such as sequences of stock market prices, both as input and output. A vector is a string containing a finite number of data points (real numbers), while a matrix is a structure composed of a finite collection of vectors, all of the same length. Even when data comes from unstructured sources, such as text or images, it is presented to the machine learning model in a structured format, often after complex human-guided transformations, such as converting words into vectors.

Transformers, the models underlying large language models, can process unstructured inputs like natural language queries, images, and more in their raw form. They can also incorporate natural language instructions and transform the input into another unstructured output, such as answering questions, translating text between languages, creating charts, or generating summaries. To learn how to perform a specific task, both transformers and traditional machine learning systems must be trained using historical data relevant to the task. It is important to note that transformers like GPT and similar models are pre-trained, just like traditional models. After pre-training, they can be fine-tuned to perform specific tasks where they process inputs and instructions, as seen in Natural Language Processing applications.

Within a company, data analysis follows a series of interconnected stages. The first stage is identifying data sources, which involves searching for and pinpointing the origins of information. These sources may include corporate databases, Customer Relationship Management (CRM) systems, web data, social media, and Internet of Things (IoT) sensors, all of which are closely tied to the company's specific operational domain. Once the sources are identified, data acquisition takes place. This process can be automated or manual, depending on the complexity of the data and the nature of the sources. The collected data is then stored and managed in dedicated structures such as databases or data warehouses, organized to facilitate easy access and prepare them for further analysis.

The next stage is data analysis, aimed at identifying patterns, trends, and relationships that can inform decision-making within the company. Finally, the results of the analysis are transformed into meaningful, actionable insights for the business. These insights can be presented through reports, dashboards, or other tools that facilitate their use. Within this process, by combining statistical and computational thinking in domain-specific problem-solving, Data Science represents the set of techniques that enable data-driven decision-making. As such, it encompasses data collection, data cleaning, data engineering, machine learning, and data visualization—many of which are also integral to AI. Data Science also aids in predictive and descriptive analysis, utilizing advanced algorithms and statistical techniques to identify patterns, understand behavior, and make forecasts. In a business context, Data Science helps optimize processes by identifying areas for efficiency improvement and cost reduction. A crucial aspect of this field is Data Visualization, the art and practice of visually representing data by converting complex information into graphs, diagrams, maps, or other easily understandable visual formats. The primary goal is to make data more accessible, comprehensible, and useful for analysis and communication.

As we have already noted, machine learning can use both labeled and unlabeled data. Labeled data includes past records with known outcomes. Examples of labeled data include banking data, where we know which customers defaulted on loans, and medical records, where we know whether a patient experienced a cardiovascular disorder. Unlabeled data consists of raw behavioral sequences. Examples include a user's purchase history on an e-commerce site or a pilot's flight control actions. Machine learning tasks can be categorized into three main types: descriptive, predictive, and prescriptive.

The *descriptive* approach focuses on analyzing past or present data to identify patterns and trends. The goal is to answer questions such as "What happened?" or "What is happening?" Through tools like reports, graphs, and statistical analyses, descriptive machine learning provides a clear picture of the current situation. An example would be analyzing a company's sales over the past 12 months to identify the most profitable periods. Another example is clustering customer profiles or patient data into groups of similar individuals, which can then be used to design targeted marketing campaigns for revenue optimization. This task typically relies on unsupervised machine learning models.

The *predictive* approach aims to answer the question, "What could happen?" By analyzing past data with machine learning techniques and statistical methods, predictive ML builds models to anticipate future outcomes. For example, a business can estimate the next quarter's sales volume, forecast customer behavior, or predict user satisfaction in a marketing campaign. Another common application is analyzing past loan applicants to predict the likelihood of default for new applicants. Predictive modeling relies on algorithms such as regression analysis, neural networks, and time-series forecasting.

The *prescriptive* approach seeks to answer the question, "What should we do?" This type of analysis suggests optimal actions to achieve a desired result. By integrating insights from descriptive and predictive models, prescriptive machine learning provides actionable recommendations, combining mathematical models, simulations, and optimization techniques. A practical example is inventory management, where prescriptive analytics helps determine the optimal number of products to order to minimize costs while ensuring availability. Another example is strategic marketing planning, where prescriptive machine learning can determine the best customer segmentation strategy to maximize conversion rates. A particularly significant application of prescriptive analytics is in autonomous driving. By analyzing the driving behaviors of millions of human drivers, machine learning can derive the best strategies for responding to environmental conditions and making driving decisions in real time.

These three approaches—descriptive, predictive, and prescriptive—are not isolated but complement each other. For instance, a company might first analyze historical sales data (descriptive), then predict future demand for its products (predictive), and finally optimize its inventory based on these forecasts (prescriptive). Together, these methods create a cycle of analysis that supports increasingly informed and strategic decision-making.

In the modern landscape of machine learning, recent advancements like OpenAI's GPT models, perform tasks that go beyond these three categories. Instead of merely analyzing past data or predicting future trends, these systems transform one type of information into another. Examples include automatic language translation, answering questions, summarizing texts, generating images from natural language descriptions, converting datasets into visual graphs, and even automatically generating computer code. These AI models take natural language text as input and produce useful outputs across a wide range of applications, demonstrating the immense potential of machine learning beyond traditional classification.

Interpretability, Explainability, and Reliability in AI

Within organizations or corporate environments, processes are monitored using appropriate sensors or data collection software, with the specific procedure varying according to the context. This generates a database, which is typically divided into two parts: a training database, used to train the model, and a validation database, used to assess the model's performance and refine it if necessary. Model training requires human intervention, as an operator defines the model's structure and selects the input features (such as parameters and hyperparameters in artificial neural networks) through an optimization process during training. Once trained, the model becomes the core of machine learning algorithms.

These algorithms are often "black boxes," meaning they operate in a way that is not easily interpretable by humans. While the decisions made by the algorithm may be highly accurate, it does not provide an explanation of how it arrived at those conclusions. Interpretable models offer insights into how knowledge is extracted from the vast amounts of data in the training database. This can lead to greater awareness and understanding for users and facilitate the generalization of the algorithms to new, previously unconsidered cases. These aspects will be further discussed later in this book.

The ability to understand and justify the decisions made by an AI system is a crucial aspect of human intelligence. It fosters trust in social and educational interactions, where understanding the reasoning behind choices is essential. The same principle applies to technical fields such as medicine, where a doctor's explanation of a treatment reassures the patient and builds trust. Similarly, understanding the decision-making process in AI systems is critical, especially when these technologies are applied in sensitive or high-stakes environments. Blindly trusting a black-box system without verifying its functioning is risky. For example, an AI system used in healthcare to predict pneumonia risk made incorrect conclusions, suggesting that asthmatic patients with heart problems had a lower risk of death than healthy individuals. This error, obvious to an experienced physician, resulted from biased data: patients with severe conditions were under close medical supervision, which reduced their mortality risk. However, this correlation did not imply causation. Such errors highlight the importance of interpreting AI models to prevent misleading data from leading to dangerous decisions.

For an AI system, *interpretability* refers to the extent to which a human can understand *how* an AI model makes its decisions, while *explainability* is the ability to describe *why* a model made a specific decision. Interpretability focuses on the model's transparency—how it works internally—whereas explainability aims to justify and describe the model's decisions, even in those cases where the inner workings are not fully interpretable.

For example, in a linear regression model, interpretability is high because it is clear how each feature contributes to the outcome through the model coefficients. Similarly, in a decision tree, you can trace the path the model follows to reach a decision. On the other hand, a deep neural network used for image classification, which is inherently complex and difficult to interpret, can still be explainable if it can highlight which parts of an image influenced the classification.

In practical applications, a loan approval AI should be able to explain why an applicant was rejected, providing insights into the decision-making process. Similarly, in a self-driving car, interpretability means understanding how each sensor and algorithm contributes to the vehicle's steering, acceleration, and braking. Explainability, however, would involve providing a clear reason why the car made a specific maneuver—such as swerving to avoid an obstacle—without necessarily revealing the full complexity of the underlying algorithm.

AI tools such as SHAP (SHapley Additive Explanations) and LIME (Local Interpretable Model-agnostic Explanations) can be used to improve the explainability of machine learning models. Both methods help clarify—at least partially—why a model made specific decisions, making the results more transparent and understandable to humans. *Explainability* is not just about transparency; it is also a crucial tool for improving AI models. Understanding a system's weaknesses is the first step in correcting and strengthening it. Interpretable models make it easier to detect biases, compare alternative approaches, and select the most suitable solution for a given need. Studies have shown that two models with similar performance can rely on entirely different decision-making criteria. A transparent analysis of these criteria allows for the selection of the best model for a specific application. The more we understand how a model operates and why it makes mistakes, the better we can improve it.

The ability to explain a model's functioning also creates opportunities for discovering new knowledge. AI systems trained on vast amounts of data can identify patterns that humans might overlook. For example, a model playing Go discovered innovative strategies later adopted by professional players. Similarly, in physics, chemistry, and biology, explainability enables the exploration of new laws and relationships that go beyond simple data-driven predictions. In these contexts, understanding the model is not just an added benefit—it is a crucial means of advancing scientific knowledge.

Furthermore, transparency has ethical and legal implications. Without transparency, it becomes difficult to determine responsibility in the event of errors. Individuals affected by AI decisions, such as a customer denied a bank loan, have the right to a clear explanation of the criteria used. This principle is so significant that the European Union introduced the "right to explanation," ensuring users can understand the algorithmic decisions that impact them. As AI becomes increasingly integrated into daily life, ensuring transparency and comprehension not only builds trust in AI tools but also promotes responsible and sustainable technology adoption.

Another defining characteristic of a machine learning model is its *reliability*, or *trustworthiness*, meaning how safe and reliable it is—an essential feature in critical applications such as healthcare, finance, and autonomous driving, where errors or opaque decisions can have serious consequences. For a model to be trustworthy, it must be accurate, meaning it can make correct and consistent predictions on new, unseen data. Its performance should be assessed using appropriate metrics to confirm its ability to solve the specific problem at hand. An algorithm's trustworthiness goes beyond the accuracy of its predictions—it encompasses the entire process of design, implementation, and use, ensuring that the system is ethical, robust, and secure. For example, in a medical setting, a reliable model must not only diagnose diseases accurately but also be interpretable, fair, and capable of operating without risk to patients. Trustworthiness is fundamental for building user confidence and mitigating the risks associated with AI adoption.

Another crucial factor is *robustness:* the algorithm must be resilient to variations or noise in input data, ensuring that small modifications do not lead to significantly different or incorrect results.

Fairness is is another key element. The model must avoid any form of discrimination arising from biases in the training data, ensuring fair and impartial decisions regardless of sensitive factors such as ethnicity, gender, or age.

The model must be protected from adversarial attacks—intentional attempts to manipulate its predictions by introducing altered inputs. An AI model must adhere to regulations such as the GDPR (General Data Protection Regulation) in Europe, ensuring transparency in automated decisions and protecting users' personal data. Last but not least, external validation is also essential: the model's performance and behavior should be verified by domain experts or through independent evaluations.

Before concluding this chapter, it is worth emphasizing that machine learning requires sufficient and high-quality data—without which effective models cannot be built. Some situations may involve data scarcity, such as underrepresented minority groups. In other cases, the training set may include numerous examples of common scenarios but lack data on rare events. For instance, datasets used to train self-driving cars may contain few examples of extraordinary situations, such as a wild animal suddenly crossing the road. Yet, even in these situations, AI must operate at peak efficiency. More abstractly, if we postulate the equivalence between data and information (the more data we have, the more informed we believe we are), information is not inherently meaningful; it depends on context. For data to be useful or even significant, it must be interpreted through the lenses of experience, domain knowledge, and sometimes even cultural and historical context. When information is contextualized, it becomes knowledge. When knowledge inspires convictions, it transforms into wisdom, and these convictions, combined with knowledge, allow people to explore new horizons.

By merging the experience encoded in data with contextualization supported by scientific knowledge, a new branch of machine learning has recently emerged: scientific machine learning. We will explore this field in Chap. 11.

6

Generative AI: The Sudden Boom

Generative Artificial Intelligence represents an advanced class of models designed to create original content that reflects the characteristics of an initial dataset. It can be described as a model capable of generating human-like text in response to a *prompt* (i.e., a command or an invitation to provide a response). Given an incomplete sentence, it can suggest a completion; given an initial phrase, it can develop entire paragraphs; given a question, it can provide plausible answers; with a topic and some basic information, it can craft an essay; given a fragment of dialogue, it can continue the conversation with a coherent transcription.

The applications of generative AI are numerous and span text, image, and sound creation. A particular type of generative AI model is that of the large language models, designed to understand and generate text in an extremely sophisticated way. These models are trained on vast amounts of textual data (hence the term *large*) to learn linguistic structures, syntax, and specific contexts. Their main characteristic is the ability to generate coherent and contextually relevant text, responding to questions, completing sentences, or creating original content.

Today, these models play a key role in raising awareness of AI's potential, largely thanks to the availability of several open-access AI systems. The following list—while certainly not exhaustive at the time of writing—illustrates the vibrancy of this sector since the launch of ChatGPT by OpenAI in November 2022. (GPT stands for Generative Pretrained Transformer.)

- *Claude* is an advanced AI tool developed by *Anthropic*, designed for natural language processing and conversational tasks. It excels in generating human-like text, answering questions, and assisting with tasks such as drafting emails, writing code, and summarizing information.
- *Llama 4*, the latest in Meta's series of open-source AI models, supports multimodal functionality, allowing it to process both text and images, enhancing its versatility for tasks like document analysis and computer vision applications.
- *Gemini* 2 is an AI tool developed by Google DeepMind, designed to push the boundaries of artificial intelligence by integrating powerful language capabilities with problem-solving skills. It combines natural language understanding, reasoning, and knowledge retrieval into a comprehensive system, by unlocking new possibilities for AI agents—intelligent systems that can use memory, reasoning, and planning to complete tasks.
- *Synthesia* is a tool designed to create high-quality video content quickly and efficiently using artificial intelligence. It allows users to generate videos with AI avatars that can speak multiple languages, making it highly versatile for global communication.
- *ElevenLabs* is an AI platform primarily known for its text-to-speech capabilities and AI voice generator. It uses deep learning models to generate highly realistic and customizable voices from written text.
- *Grok 3* is an AI-based tool developed by xAI to help users better interact with complex information, enabling a faster and deeper understanding of various topics. The term "grok," originally from science fiction, refers to the ability to fully and intuitively understand something. Grok 3 can analyze images and respond to questions, and powers several features on Musk's social network, X.
- *Mistral AI* is a relatively new but innovative company in the field of artificial intelligence, focusing primarily on building advanced open-source AI models and tools. The company has gained prominence as an alternative to proprietary AI systems as it aims to "democratize" AI by focusing on open-source innovation.
- *CopyAI* is an AI-powered writing tool designed to assist users in efficiently creating high-quality content. Using advanced natural language processing models, CopyAI generates text for a wide range of purposes, including marketing copy, blog posts, social media captions, email templates, and more.

Transformers and Self-Attention Mechanisms: The Engines of Generative AI

Large language models rely on transformer algorithms (previously mentioned), which use the attention mechanism—a concept that can be imagined as the ability to focus on key words while reading a sentence. This process, represented in Fig. 6.1, consists of layers of transformers and neural networks, effectively capturing complex relationships in natural language. The weighted information is combined to create a representation of the context. This approach has proven useful, for example, in machine translation and other natural language processing tasks.

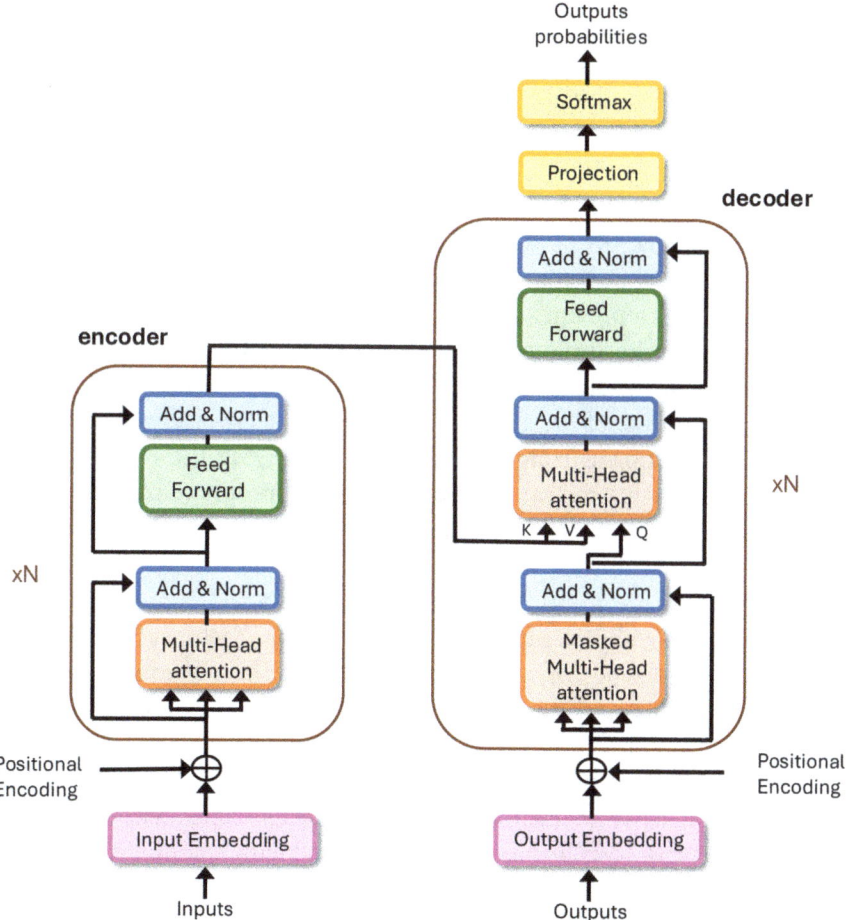

Fig. 6.1 The transformer

The algorithm behind a large language model, such as ChatGPT-4, is structured in several steps. The first step is to convert the text into *tokens*, linguistic units that can be words or parts of words (subwords). This tokenization process converts the text into a sequence of integers (an index for each token in the model's vocabulary). Example: The text "The cat is sleeping" could be transformed into the tokens [1001, 459, 34, 8873]. Each token is then transformed into a numerical vector via a so-called embedding matrix. This matrix assigns each token a dense vector representation in a continuous space so that semantically similar tokens have nearby vectors. For example, [1001, 459, 34, 8873] becomes four embedding vectors, each of a fixed size, such as 768 dimensions.

Since the transformer does not have an integrated sequential mechanism like recurrent neural networks (RNNs), positional encoding is added. This technique allows the model to know the order of the tokens in the sequence. The formula for positional encoding is defined using trigonometric functions, sinusoidal and cosinusoidal, which add a small amount of variation to the embedding vectors depending on the position of the tokens.

The core of the algorithm is the self-attention mechanism. Here, the model evaluates the relationship between each token and all the other tokens in the sequence, creating a sort of map of how much each word "attracts" or "influences" the others. Mathematically, for each token, three matrices are computed: Q (Query) represents the current word, K (Key) represents the other words with which the current word can interact, and V (Value) represents the informational content associated with each word. Self-attention is computed by taking the product of the Q and K matrices (which measures the similarity between tokens), normalizing it, and then multiplying the result by the V matrix (which weights the values to produce the appropriate context).

To improve the model's ability to capture different types of contextual relationships, the Transformer uses the so-called multi-head attention, an algorithm that performs multiple parallel self-attention computations (called "heads"), each with a different projection of the data. Each head specializes in capturing specific types of relationships, such as the relationship between nearby words or between distant words. The results of the different heads are then concatenated together.

After the self-attention mechanism, the result is passed through a feed-forward neural network. Each token is processed independently by a fully connected network (two linear layers interspersed with a nonlinear activation function, such as *ReLU*).

The transformer consists of many layers of multi-head attention and feed-forward networks applied sequentially. Each layer adds further processing to the context of the text. The output of each layer is normalized via a technique called layer normalization. After the text has been processed through all the layers of the Transformer, a refined numerical representation for each token is obtained. This representation is then passed through a softmax function, which converts the result into a probability distribution over the model's vocabulary.

The *softmax function* is defined as

$$\mathit{softmax}(z_i) = \frac{e^{z_i}}{\sum_j e^{z_j}}$$

where z_i is the logit associated with token i. (In statistics, the logit model, also known as the logistic model or logistic regression, is a nonlinear regression model used when the dependent variable is dichotomous, meaning it can take only two distinct values. The goal of the model is to determine the probability that an observation will produce one or the other value of the dependent variable. It is worth noting that logits, although they are probability indicators, can also take on negative values. It is the *softmax* function that transforms them into all positive values (thanks to the application of the exponential function) and normalizes them.) The token with the highest probability is selected as the prediction. The algorithm can generate one token at a time in an autoregressive mode: once a token is generated, it is added to the context, and the process repeats to generate the next token, until a stop condition is reached (such as an end-of-sequence token).

The model is trained using a massive amount of textual data, minimizing a cost function like cross-entropy, which measures the difference between the predicted probability for a token and the target distribution (the correct token). The model weights (i.e., the parameters **W** and **b** encountered in Chap. 4) are updated via optimization algorithms such as the Adam algorithm (Quarteroni et al. 2025).

We can summarize the entire process by outlining its distinct phases:

Tokenization: the text is divided into tokens; *Embedding*: tokens are transformed into numerical vectors; *Positional Encoding*: information about the position of the tokens is added; *Self-Attention*: the relationships between tokens are calculated; *Multi-Head Attention*: multiple self-attention heads work in parallel; *Feed-Forward Network*: tokens are processed by neural networks; *Layer Repetition*: many layers of processing refine the context;

Softmax: the output is converted into probabilities; *Prediction*: the next token is generated based on the probabilities.

We can now ask how the tokens: 1001, 459, 34, 8873 are assigned to the text: "the cat is sleeping." They depend on the model's vocabulary, a predefined list of words, parts of words (subwords), or characters, where each element corresponds to a unique number called the token index. A model like GPT uses a vocabulary built through a tokenization technique called Byte Pair Encoding, which splits the text into basic units such as common words, subwords, or even individual characters, to maximize efficiency. In this case, "The" could be represented by the token 1001, "cat" could be represented by the token 459, "is" could be represented by the token 34, "sleeping" could be represented by the token 8873. The model has a pre-trained vocabulary that associates each word or part of a word with a number. During training, the vocabulary is created by analyzing huge amounts of text, and each token is assigned an index based on its frequency and usefulness.

By splitting words into pieces or subwords, Byte Pair Encoding allows the handling of rare or unknown words (or simply misspelled words!) that are not present in the vocabulary as wholes. For example, for "sleeping," "sleep" could be a common token and have its own index, such as 5123, while "ing" could be another frequent token, with a separate index, such as 4872. In this case, "sleeping" could be tokenized into two parts: [5123, 4872]. The model joins these tokens to generate the full meaning. In general, this process allows models like GPT to handle a wide range of words, even new or rare ones, by breaking words into known pieces (subwords), which is especially useful in languages with complex morphology or with infrequent technical terms.

The development of large language models is progressing rapidly, with new releases continuously adding new capabilities. The latest models are multimodal, meaning they can integrate different types of sources, such as text, audio, images, and video. Conventional language models can be specialized to perform different common tasks, such as sentiment analysis and named entity recognition, without requiring additional specialized knowledge. However, to address more complex and content-rich tasks, systems based on language models can be implemented that draw on external sources of knowledge. This ensures greater consistency in facts, improving the reliability of the generated answers and reducing the risk of errors.

The Hallucinations of Generative Intelligence

Hallucinations in generative artificial intelligence occur when an AI model, like ChatGPT, generates responses or content that appear coherent and plausible but are completely fabricated, incorrect, or not based on real facts. Although the term may not have been the best choice, this phenomenon happens because generative AI does not have a real understanding of the world or the information it processes. Generative AI algorithms are designed to produce responses that seem plausible and linguistically coherent but are not necessarily factually accurate. Since the model optimizes for the probability of the next piece of text occurring rather than accuracy, it can generate wrong answers, even if they seem plausible, stylistically correct, or contextually appropriate.

The causes that lead to hallucinations can vary. Some of the most common ones include: ambiguity or lack of clarity in the input (i.e., the prompt); lack of information in the training set; biases in the training data; the model's structure, which can lead to incorrect correlations between input and output, producing responses that seem logical but are not; training techniques, which program the model to generate a response even when it does not have complete or precise knowledge (these models never manage to say "I don't know!"); and the limitations of the model's memory, which means it cannot recall previous information during long conversations.

In generative AI models, like those based on transformer neural networks (e.g., GPT), a parameter called *temperature* controls the randomness or variety of the responses generated by the model. It modifies the probability distribution of the words the model can choose from to generate the next piece of text. In simple terms: when the temperature is set to a low value (close to 0), the model becomes more deterministic and tends to choose the words with the highest probability. This means the model is more likely to generate predictable and safe answers, reducing creativity but increasing consistency and repeatability of responses. When the temperature is higher (above 1), the model introduces more randomness into the word selection process. With a higher temperature, the model may choose less probable words, leading to more varied and creative responses, but also potentially less coherent or riskier ones. In summary, the temperature is a key tool for balancing creativity and consistency in generative AI models, allowing users to control how "adventurous" or "conservative" the model should be in generating text.

To mitigate hallucinations, developers and researchers are constantly working to improve the coherence and accuracy of large language models. This includes data filtering operations, changes to the model's architecture, and advanced training techniques such as reinforcement learning from human feedback (RLHF), which provides feedback to the system on the number of hallucinations produced, helping it to correct itself. However, it is important to note that the phenomenon of hallucinations is intrinsic to the architecture and training process of generative AI models. Reducing hallucinations might limit the model's creativity in content generation, making it necessary to find a balance between accuracy and creativity depending on the model's use case.

To reduce issues caused by hallucinations an approach called *Retrieval-Augmented Generation (RAG)* has been introduced to handle tasks that require significant amounts of knowledge (Lewis et al. 2020; Gao et al. 2023). RAG combines an information retrieval component with a text generation model. This system is flexible and allows for continuous updates, ensuring access to the most up-to-date information without having to start from scratch. RAG receives input and retrieves relevant documents from external sources, such as Wikipedia or those found in the archives or information systems of a specific company, using them as context alongside the original input to feed the text generator and produce the final output. This makes retrieval augmented generation models adaptable in situations where information evolves over time, providing always up-to-date responses and forcing it to adhere as closely as possible to the informational context of relevant documents. Creating a retrieval augmented generation application, such as a chatbot, involves integrating techniques based on information retrieval with generative models. The process begins with data preparation, during which a large domain-specific corpus of information is collected, pre-processed, and organized into a document archive. The next step is indexing, which involves representing the documents as dense vectors that are then indexed in a vector database to allow for efficient similarity searches. When a user submits a query, it is processed by a retrieval module that identifies the most relevant documents in the database. The retrieved documents are then provided as context to a generative model, such as GPT-4, LlamA, or others. This model generates a response by combining the context of the documents with the user's query, often formatted into a structured prompt. The retrieval module and the generator are integrated into an API or cohesive application layer. (An API – *Application Programming Interface* – is a set of rules and tools that allow different applications or systems to communicate with each other, like a bridge that allows a software to access the functionality or data of another software without having to know the internal details.) Finally, the retrieval augmented generation pipeline is deployed on cloud platforms. Through these stages, the retrieval augmented generation system becomes a robust and scalable solution for intelligent chatbot applications.

Open Problems That Deserve Attention

Despite the great potential of generative AI in creative innovation and generating unique content, there are many issues to consider. First, we are used to computing tools like email, search engines, and programming languages that are rigid and not creative, but reliable and predictable. Now, however, we have language models like ChatGPT that understand natural language requests and generate creative images and texts, but without the same reliability, predictability, or accuracy—indeed, they are prone to hallucination. How can we use a powerful but unreliable tool? For applications where errors can cause real damage (for example, in medicine, to provide therapeutic guidelines, or in business, to make financial decisions), caution is necessary. We also cannot ignore the risks of creating false or manipulated content. Therefore, the evolution and responsible adoption of these technologies require careful consideration of ethical and security aspects.

Another aspect concerns the credibility and reliability of the data used to train generative AI. The rise of generative AI is rapidly filling the internet with synthetic content (i.e., non-original content produced by the AI itself). At the time of writing, for example, Newsguard has identified over six hundred news and information sites created by AI systems. In the future, AI models could be trained primarily on content generated by generative AI itself! This self-feeding process seems destined to create a sort of informational echo within the AI itself. Recent studies have shown that training models on data produced by other AI models leads to a degenerative and irreversible process: the model ends up overestimating the most likely events and underestimating rare ones, progressively losing touch with the original data distribution. This phenomenon, called "model collapse," occurs in different types of models, from large language models to image generators, and only a few generations of training are enough to compromise the quality of the results. Another study documented a similar process, called *Model Autophagy Disorder*, which occurs when generative models do not receive a constant stream of real data. Without a continuous flow of human-generated content, generative AI models risk deteriorating. This prospect of an authenticity crisis could jeopardize the value of generative AIs, especially considering that many consumers are skeptical about content created by algorithms. The very essence of the internet is based on interaction and human-generated content: the invasion of artificial content risks undermining the intrinsic value of the web.

In this new digital landscape, where authentically human content is increasingly rare, a reconsideration of incentives and economic models that value human contributions seems necessary. The question is: how can we create an ecosystem that encourages people to continue producing and sharing authentic content? To do this, we need to focus on three fundamental aspects: fair compensation for human creativity, greater control over creative results through transparency, and technological innovations that clearly distinguish human-generated content from machine-generated content. This is certainly not a simple program to implement.

There are several other fundamental aspects that arise in the field of machine learning. How can we program systems that learn automatically from data and improve their predictive and prescriptive abilities with experience? Learning means giving meaning to a subject, event, or emotion, interpreting it through our words or actions. It involves using new skills or knowledge, integrating them with the skills and understanding we already possess, to apply them in a useful way: gathering available data (ingestion), acquiring knowledge (understanding and interpreting data, turning it into knowledge), and finally using that knowledge to take concrete actions (i.e., acting). But how do humans learn? Certainly through teaching and observation, but we also learn by doing and making mistakes (trial and error). Fundamentally, machine learning systems learn in one of these two ways. There are things humans cannot learn (but perhaps machines can) and others that are difficult to teach to machines (one of many: common sense).

For humans, learning is difficult when there is too much data, as they cannot easily make sense of it, such as finding regularities in the human genome, learning to recognize an object among millions, or analyzing and predicting how the market will evolve. When data is highly heterogeneous and changes too frequently, or when operating in difficult or dangerous environments where the trial-and-error strategy is not advisable, learning becomes challenging. It is often difficult for humans to teach as well when there is a lack of theoretical knowledge and not enough prior information or experience to "understand and acquire knowledge." For example, we still do not fully understand the process through which our brain recognizes images and language.

If properly trained, for some applications and to some extent, machines are better than humans: in games, in performing precision surgery, in image analysis and recognition, and, recently, in natural language recognition. However, even in these cases, the main limitation is the explainability of the underlying algorithms. They are "black boxes," while, for example, mathematical models representing real-world processes are "white boxes," completely interpretable.

Large language models are characterized by an extraordinarily high number of trainable parameters. GPT-3, for example, had 175 billion parameters, while the subsequent GPT-4 version, which extended its capabilities to image generation, is estimated to have over a trillion parameters. This vast number of parameters has various implications in terms of costs, training times, access to necessary data, and other operational costs. As for tokens, which are manageable text fragments (words, characters, or parts of words) that the model can use, the number has increased from 4096 in GPT-3 to 32,768 in the extended version of GPT-4. A larger number of tokens allows for analyzing longer texts or maintaining complex contexts in conversations, making the models more useful in advanced applications.

Accessing quality and adequately sized datasets can be challenging, especially considering issues related to data privacy. Executing and evaluating large language models also incurs significant costs in terms of computational resources and time. The model's size can make inferences slower, requiring more powerful hardware to maintain acceptable performance. Sam Altman, CEO of OpenAI, revealed that the cost to train GPT-4 was around 100 million dollars.

Sustainable? Not Quite

The use of powerful and specialized computing infrastructures for training large language models can be prohibitive for many organizations due to the associated costs. The enormous energy consumption also raises concerns about its environmental impact. According to a recent study, the two tech giants Microsoft and Google used 24 TWh in 2023. In the same year, Italy consumed around 300 TWh to meet the energy needs of over 60 million citizens. To provide another reference, Nigeria, with nearly 220 million people, consumed 32 TWh in 2023. These extremely high consumption rates are due to the complex calculations carried out by global data centers, which, according to the International Energy Agency, consumed around 460 TWh in 2022, a figure that could reach 1000 TWh by 2026. This increase is largely attributed to the rapid and intensive development of generative artificial intelligence, which requires a lot of energy, especially during training phases. At the beginning of 2025, Meta announced a Request for Proposals (RFP) to identify nuclear energy developers, with the goal of adding 1–4 gigawatts (GW) of nuclear capacity in the United States by the early 2030s. This initiative aims to support Meta's AI innovation and sustainability goals by integrating an energy source into the expanding power grid, essential for its data centers and surrounding communities. Unlike renewable sources like solar and wind, nuclear projects require longer timelines, higher capital investments, more complex regulatory oversight, and strategic planning for their longer operational cycle. Involving partners early in multiple projects is seen by Meta as key to scaling implementations effectively, reducing costs, and ensuring accurate execution. However, it is worth noting that globally, 440 nuclear plants generate about 10% of electricity. Ten times as many would be needed to meet the total electricity demand. Without a different strategy to mitigate the expected growth of data centers, this does not seem to be the (only) path forward.

If a global ranking were made on electricity consumption, including big tech companies along with states, Google and Microsoft would rank above more than 100 countries—more than Tunisia or Jordan, slightly less than Slovakia or Ecuador, or equal to Azerbaijan, which has more than ten million inhabitants. These figures come from an analysis by industry expert Michael Thomas, based on data from the U.S. Energy Information Administration (EIA) and the environmental reports of the two companies. A recent alert comes from a report in the *Financial Times*, which discussed the situation in "data center valley," Virginia, where the consumption of "blue gold" (water!) has increased by nearly two-thirds since 2019. Virginia is home to one of the largest concentrations of data centers in the world, from Amazon to Google and Microsoft. Machines used 7 billion liters of water in 2023, according to documents requested and obtained by the *Financial Times*. In gallons, it has risen from 1.13 billion pre-COVID to 1.85 billion. Environmentalists foresee a future "explosion" of thirst, specifically due to AI. According to estimates by the Dgtl Infra research group, U.S. data centers consumed over 75 billion gallons of water in 2023—enough to quench London's thirst for 4 months. The paradox raised by environmentalists is that there are areas of the same U.S. state suffering from drought. The expansion of these facilities (which is expected to double from 2019, with new areas still under construction) will only worsen this situation. In November 2023, Bank of America estimated that data centers were the tenth-largest water consumer in the U.S. Major tech companies are trying to limit their water withdrawals, but their reports highlight the difficulty of the situation: Google increased its water consumption by 14% in 2023, due to the needs of data centers. The search engine explains that 15% of its water withdrawals occur in areas with high scarcity, while Microsoft states that 42% of its global consumption is in water-stressed areas. Europe also seems to be no exception. For example, by the end of 2025, 83 new data centers are expected to be built in Italy. Only Microsoft has announced a €4.3 billion investment in data centers in Italy, along with €3.2 billion in Germany and €2 billion in Spain.

From an ethical standpoint, the limited access to financial and technological resources to train and use large language models is likely to generate increasing inequalities. Only a few large organizations or companies can afford to develop and use such models, while small businesses and independent researchers might be excluded from these advanced technologies. To mitigate these issues, it is essential to promote the responsible sharing of pre-trained models and access to computing resources through collaborative initiatives. The scientific and industrial communities should ideally work to make resources and data available, reducing the gap in access to these technologies and ensuring that they are developed and used responsibly and inclusively.

OpenAI, But Not Only

We conclude this chapter on generative AI with a historical context, given the impact that ChatGPT has had on shaping society's perception of AI. ChatGPT is the chatbot that started the race toward generative artificial intelligence. With its launch on November 30, 2022, these tools moved from research labs to our devices, bypassing the intermediary of a search window or application—an important step that generated enormous interest and widespread adoption. The more they are used, the more they train themselves and improve their responses. In addition to ChatGPT, OpenAI has developed DALL-E, a text-to-image model that generates images based on textual descriptions. Although OpenAI's ultimate goal is to create AGI (artificial general intelligence), an AI capable of reasoning and solving problems like a human, as we have seen many times, this remains a dream for now.

OpenAI's history, though recent, provides a paradigmatic example of the accelerated evolution of new companies in the generative AI field. Open AI was founded in 2015, by prominent figures such as Elon Musk and Sam Altman, aimed at promoting the development of artificial intelligence in a transparent and safe way for humanity. Originally a non-profit organization focused on technological research, in 2019 it opened a for-profit division to attract investments, the largest of which, $13 billion, came from Microsoft. Following the first language model GPT-3.5, on May 13, 2024, OpenAI announced GPT-4o ("o" for "omni"), made available for free to all users, who, by that summer, exceeded 180 million. Over the years, many of the founders left the organization, and the original concept of "openness" was gradually abandoned as the company's governance changed. The launch of ChatGPT in November 2022 marked a turning point: the global success of this model led to exponential growth for the organization, with significant resource increases. However, this rapid growth also brought strategic challenges. OpenAI has maintained a focus on creating advanced intelligences, leaning more towards the development of superintelligences than toward concrete commercial solutions. This allowed other companies, such as Anthropic, to gain ground by offering more business-oriented alternatives.

In 2023, OpenAI faced a severe internal crisis, culminating in the temporary dismissal of Sam Altman by the board of directors, led by co-founder Ilya Sutskever. This episode reflected deep tensions within the company, caused by the difficulty of balancing AI safety efforts with the need to monetize the developed technologies. The situation was resolved with Altman's return, supported by Microsoft, and Sutskever's departure, who would go on to found a new startup called Safe Superintelligence (SSI). Alongside these internal developments, OpenAI had to contend with growing concerns about the safety of its technologies. New "o1" models, capable of advanced reasoning, raised fears of their potential use in dangerous contexts, such as in warfare with the creation of biological or chemical weapons.

Another critical challenge concerns access to the data necessary to train AI models. The growing value of data has driven many holders, such as publishers and media outlets, to limit access or sell them at high prices. OpenAI has already made costly agreements, such as the one with News Corp, but the scarcity of quality data could slow future development. Looking ahead, OpenAI faces crucial decisions. The company must find a balance between the need to meet market demands and maintain an economically sustainable strategy, without compromising the foundational values that guided its creation. The future of the company will depend on its ability to win these challenges while keeping innovation and safety as top priorities. Globally, this is leading to increased regulation of the sector, with laws on the way in California and the European Union aimed at better managing the risks associated with advanced AI.

In the ever-evolving landscape of generative AI, it is essential to remain vigilant, because many emerging, still little-known companies could soon establish themselves as the new OpenAI. A significant example is DeepSeek, a Chinese start-up that is rapidly gaining ground. On January 20, 2025, it released DeepSeek-R1, a partially open-source 'reasoning' model, capable of solving scientific problems with a quality comparable to that of o1, OpenAI's most advanced language model, but with drastically lower costs and computational requirements. Only a few days later, it launched Janus-Pro-7B, a model capable of generating images from text descriptions, similar to OpenAI's DALL-E 3 and Stable Diffusion from Stability AI. DeepSeek optimized its hardware by using 2000 NVIDIA H800 GPUs, less advanced than the US H100s, but compensated for the technological gap with specific frameworks to maximize efficiency. This approach has allowed it to offer APIs at costs up to 90% lower than OpenAI. (An API—Application Programming Interface—is a set of rules and tools that allow different applications or systems to communicate with each other, like a bridge that allows a software to access the functionality or data of another software without having to know the internal details.) Just think that the price per million input tokens is only 0.14 dollars, compared to 15 for OpenAI, or a ratio of 1 to 100! This difference in cost is explained by two main factors. On the one hand, the use of *distillation techniques*, which allow knowledge to be transferred from larger models to smaller and more efficient versions, reducing the computational requirement. On the other, the partnership with AMD (Advanced Micro Devices), a semiconductor company that offers high-performance computing solutions at lower costs than NVIDIA. The impact on the market was immediate: NVIDIA stock lost 17% in a single day of trading, with a capitalization drop of almost 600 billion dollars.

Behind this rise is Liang Wenfeng, a 39-year-old entrepreneur with a degree in computer science from Zhejiang University, one of the most prestigious in Hangzhou, China. After co-founding the hedge fund High-Flyer, in 2023 he founded DeepSeek, initially with the aim of applying artificial intelligence to financial trading. Generative AI, in fact, was only a secondary project, and OpenAI was not yet a real competitor. However, the company has grown rapidly, thanks to advanced training techniques such as pure reinforcement learning, the *Mixture-of-Experts* architecture, an advanced neural network architecture that improves the efficiency and performance of AI models while reducing the consumption of computational resources, the *Multi-Head Latent Attention* technique to improve the processing of complex data, as well as the use of open-source models such as Janus-Pro and R1, which have allowed a reduction in development costs: the training of the basic model cost just 5.6 million dollars. Instead of activating the entire network for each input, mixture-of-experts selects only a few specialized "experts," optimizing the calculations. This idea is inspired by the functioning of the human brain, where different areas are activated based on the task at hand.

While for many international observers the success of DeepSeek was surprising, Chinese experts consider it a predictable result, the result of huge government investments in artificial intelligence. According to Yunji Chen, an expert at the Chinese Academy of Sciences, the emergence of startups like DeepSeek is the natural result of massive funding in the sector and the abundance of highly specialized AI researchers. The Chinese government, in fact, has declared artificial intelligence a strategic priority: in 2017, it announced the plan to become a world leader in the sector by 2030, setting 2025 as an intermediate goal to reach a globally competitive technological level. Already in 2022, the Ministry of Education had authorized 440 universities to offer AI degree courses and, in the same year, China trained almost half of the world's leading researchers in the sector, while the United States accounted for only 18%. In addition to public funding, China has had to deal with restrictions imposed by the United States on the import of advanced chips, such as the latest generation NVIDIA GPUs. To get around these limitations, DeepSeek stockpiled 10,000 A100 chips ahead of the 2023 ban and developed techniques to parallelize training algorithms to take full advantage of less powerful hardware. This focus on optimization has led Chinese research to focus on model compression and energy efficiency, strategies that have made DeepSeek particularly competitive. From a business perspective, DeepSeek has adopted a hybrid model, combining open-source and proprietary solutions. Some models, such as Janus-Pro and R1-Distill, have been released under the MIT open-source license, allowing the Hugging Face platform to launch the Open R1 project to replicate DeepSeek's training pipeline and accelerate its global deployment. However, other models remain proprietary, giving the company a competitive advantage in the market.

There are, however, some critical considerations. DeepSeek models include filters to avoid content critical of the Chinese Communist Party, an aspect that could limit their adoption internationally. Additionally, to expand into Europe, the company will have to deal with the strict regulations of the GDPR, particularly regarding algorithmic transparency and the management of sensitive data, an area where competitors like Anthropic appear to be ahead of the curve.

Meanwhile, competition is heating up. On January 29, 2025, Alibaba unveiled its most advanced language model, Qwen2.5-Max, claiming it outperforms DeepSeek V3, the large language model released in December 2024. Around the same time, Moonshot AI and ByteDance launched their own reasoning models, Kimi 1.5 and 1.5-Pro, claiming they outperform o1 in some tests. DeepSeek's impact extends far beyond China. Its innovative approach demonstrates that algorithmic efficiency can make up for a lack of advanced hardware, challenging the prevailing idea that more GPUs equal better AI. While its use of 2000 NVIDIA H800 GPUs may seem modest compared to big tech's resources, it represents an alternative to the so-called "scaling law," paving the way for a democratization of AI. (Scaling laws in AI describe the empirical relationship between a model's performance and the scale of resources used for training, including model size (parameters), dataset size (training data), and compute power (FLOPs). As these factors increase, performance (measured by accuracy, loss, etc.) generally improves.) However, the company's future will depend on its ability to maintain its innovative momentum, overcome hardware limitations and political restrictions, and, who knows, perhaps shift its focus from the simple size of models to data quality and intelligent architectures.

7

AI Is Not Just Generative

Generative AI has captured significant attention thanks to its popular successes, such as conversational chatbots (e.g., ChatGPT), image and filmgenerators (e.g., DALL-E and MidJourney), and creative tools that generate text, music, or art. However, this hype risks overshadowing the significant and long-standing contributions of traditional, non-generative AI, which has had an equally, if not more, profound impact in many crucial areas of society.

There are fundamental differences between generative AI and, say, traditional AI. Generative AI is designed to create new content—texts, images, audio, and videos—based on deep learning models trained on large datasets. It is at the heart of many recent applications that have gained widespread media attention, primarily due to their ability to imitate human dialogue and creativity and directly engage users. Tools like chatbots and image generators allow direct interaction with users, quickly capturing their imagination. Their ability to produce original content, such as poems, articles, or artworks, makes them fascinating and accessible. Advancements in generative AI have emerged rapidly, surprising the public and demonstrating clear progress in how machines can generate outputs that were once exclusively human.

A further extension of the (already vast) field of language models is that of so-called AI agents. These are expert AI systems designed for specific domains, software programmed to perform tasks autonomously, often replicating human decision-making processes within specific contexts. These systems "perceive" their environment, process received information, and take actions aimed at achieving predefined goals, leveraging advanced technologies such as machine learning, natural language processing, and computer vision. AI agents operate without constant human intervention, using sensors, APIs (Application Programming Interfaces), or data inputs to perceive their surroundings. An example is a virtual assistant capable of understanding user requests through natural language. They process received data, make decisions, and act based on preset models or predictions derived from machine learning algorithms. Some agents even continuously improve through reinforcement learning, meaning they learn from data and feedback.

Designed to pursue specific objectives, AI agents can answer questions, optimize processes, navigate physical spaces, and interact with users. Their design varies: some are simple and reactive, merely responding to stimuli without memory, while others are more sophisticated and proactive, capable of planning and evaluating the consequences of their actions. While agents like drones and autonomous delivery robots operate in the physical world, others function solely in digital environments. These agents are applied in various sectors: from chatbots and virtual customer service assistants to diagnostic tools in healthcare, financial trading systems, fraud detection, intelligent video game characters, and logistical solutions for route optimization. As technology evolves, AI agents are becoming increasingly complex and integrated into different aspects of society. They play a crucial role in industries and daily life, serving as fundamental tools for creating intelligent, flexible systems that interact effectively with both their environment and people.

Creativity at Your Fingertips

With just a few keystrokes, anyone can ask a generative AI program to write a novel, a rap song, or a theatrical piece. While these models can produce human-like writing in mere seconds, some researchers argue that large language models do not create anything truly new. Critics claim they are merely "stochastic parrots" that blindly remix the words they were trained on. Expecting the originality of William Shakespeare or Dante Alighieri would be excessive. Indeed, the debate over how "creative" generative AI really remains highly open. Testing this creativity "objectively" is challenging. To address this, researchers have developed metrics to analyze both the absence of plagiarism and the human perception of factors such as fluency and originality. A program called DJ Search, developed at the University of Washington, compares AI-generated text with vast online databases. This tool does not merely look for identical matches but also analyzes sequences of words with similar meanings, assigning a quantitative "significance index." Comparing the linguistic novelty of published novels, poems, and speeches with texts written by the latest generation of large language models, researchers found that human authors outperformed AI by about 80% in poetry and 100% in novels (Lu et al. 2024).

Of course, these results should not be taken as absolute, as they are influenced by subjective criteria. However, it can be cautiously affirmed that, as of today, the creativity of AI language models is nowhere near that of real human authors! That said, AI language models can still serve as "assistants" to humans, generating first drafts that can later be refined with creativity, inspiration, and originality.

There Is Life Beyond Generative AI!

As we have observed, the hype around generative AI risks diverting attention from the critical contributions of traditional AI. While content generation is impressive, many vital functions in modern society are supported by non-generative AI systems that remain essential for the operation of our infrastructures.

Traditional AI includes supervised learning algorithms, recommendation systems, classical neural networks, decision trees, and optimization systems. It focuses on automating processes, classifying data, or predicting events. Often less visible to the public, it is widely present across multiple sectors. Here are some of its numerous successes (note: this list is not exhaustive!):

- *Healthcare*: Non-generative AI has long been used for facial recognition, medical data analysis, and disease diagnosis (e.g., cancer detection using convolutional neural networks (CNNs) to analyze radiographic images and identify anomalies). Decision-support systems help doctors make more informed choices. AI also optimizes healthcare resource management, such as hospital bed allocation and medical supply distribution.
- *E-commerce and Streaming*: Recommendation systems on platforms like Amazon or Netflix rely on traditional AI, using techniques such as collaborative filtering and regression models. These systems have had a massive economic impact, driving sales and enhancing user experience.
- *Finance and Fraud Detection*: Traditional AI has been used for years in banking fraud detection and credit card transaction monitoring. Machine learning systems classify thousands of transactions in real time, identifying suspicious ones. Many non-generative models also support financial trading by analyzing vast amounts of data and making rapid decisions with significant market impacts.

- *Traffic and Logistics*: AI manages city traffic, optimizes real-time routes, and improves road safety. Many companies use traditional AI to optimize delivery routes, reducing costs and emissions.
- *Manufacturing and Robotics*: Traditional AI powers industrial automation, guiding robotic arms, quality control systems, and automated production lines. Robotics has revolutionized manufacturing, improving efficiency and reducing costs. The term "robot," derived from the Slavic word for "forced labor," was coined by Czech writer Karel Čapek in the 1920 play *R.U.R.* (Rossum's Universal Robots), where robots were humanoid machines made of organic material designed to ease human labor. In the 1940s, American writer Isaac Asimov redefined the concept, transforming robots into mechanical artifacts. By the 1980s, robotics was described as the science of intelligent connections between perception and action. Today, machine learning in robotics enables functions like object grasping, object categorization, and even linguistic interaction with humans. Learning can occur through self-exploration or human guidance. For example, artificial vision allows robots to perceive, identify, and navigate their environment using machine learning algorithms and sensors. These methods are widely applied in production procedures such as material inspection and pattern recognition. One particularly advanced form of learning, self-supervised learning, allows models to recognize patterns from data without explicit human-provided labels. This enables robots to learn tasks autonomously and adapt to changing environments (Siciliano 2019).
- *Autonomous Vehicles*: A prominent example of non-generative AI success is autonomous vehicles. Many car manufacturers have developed prototypes capable of driving at moderate speeds on busy roads. These cars use an array of sensors, lidars, cameras, and localization systems such as GPS, applying machine learning algorithms to make real-time decisions about braking, sudden direction changes, and different driving modes. Despite significant progress, widespread autonomous driving is not imminent—not only due to technological hurdles but also because of legal challenges, as there are currently no universal regulations governing its use on public roads. However, at the time of writing, thanks to a partnership between Google's Waymo and Uber, it is already possible to book self-driving taxis in a few U.S. cities.

In conclusion, generative AI and traditional AI are not rivals but complementary forces. While generative AI enhances accessibility and creativity, traditional AI remains deeply embedded in sectors where precision, robustness, and reliability are paramount. Recognizing and valuing both paradigms is essential to understanding how AI will continue to drive innovation and improve society in countless ways.

8

AI in Science

In fields where the output of artificial intelligence can be independently verified, AI has already demonstrated its extraordinary effectiveness in various areas of research, both in science in general and in mathematics.

A Nobel-Worthy AI

In October 2024, the Nobel Committees in Stockholm awarded the Nobel Prizes in Physics and Chemistry for work related to artificial intelligence, marking a historic moment in science. The Nobel Prize in Physics was awarded to John J. Hopfield and Geoffrey E. Hinton for "fundamental discoveries and inventions that enabled machine learning with artificial neural networks." The Nobel Prize in Chemistry was awarded in part to David Baker for "computational protein design" and in part to Demis Hassabis and John M. Jumper for "protein structure prediction." These historic recognitions highlight AI's contribution to science, medicine, and society, celebrating the winners' discoveries and their impact on the future of human health and knowledge. Hopfield and Hinton developed the technological foundations that support modern machine learning. In 1982, Hopfield created the *Hopfield network*, an associative memory structure capable of storing and recalling information (Hopfield 1982). In 1985, Hinton developed *the Boltzmann machine* (see, e.g., the article Salakhutdinov and Hinton 2009), a method that allows AI systems to autonomously discover patterns in data. More precisely, Hopfield developed a neural network capable of storing and reconstructing patterns.

The network's nodes can be visualized as pixels, and its functioning is inspired by the physics of atomic spin—a property that makes each atom behave like a tiny magnet. The Hopfield network is mathematically described using an energy model like that of spin systems in physics. During training, connection values between nodes are adjusted so that stored images correspond to low-energy states. When presented with a distorted or incomplete image, the network iteratively updates the node values to minimize its energy, gradually converging toward the closest stored image. Drawing from statistical physics—the study of systems composed of many interacting components—Hinton designed the Boltzmann machine to be trained by exposing it to representative examples of the patterns it seeks to model. Once trained, it can classify images or generate new ones resembling those in its training set. Hinton's advancements in neural networks laid the groundwork for the modern surge in machine learning research.

These contributions paved the way for artificial neural networks, which now power technologies capable of analyzing vast amounts of data and learning efficiently. Thanks to these innovations, computers can now interpret images, make predictions, and interact naturally with humans. Tools like OpenAI's ChatGPT are made possible also because of Hopfield and Hinton's fundamental contributions. Their discoveries have revolutionized key sectors, including digital medicine, and represent a technological milestone, comparable by some to the invention of the printing press or the internet.

Hassabis and Jumper successfully tackled one of modern science's most complex challenges: predicting the three-dimensional structures of proteins based on their amino-acid sequences. Their model uses AI to generate highly accurate predictions, solving a problem that had hindered scientific progress for decades. In 2010, Hassabis co-founded *DeepMind*, an AI company later acquired by Google in 2014. In 2018, he developed *AlphaFold*, which enabled protein structure predictions with about 60% accuracy—20 percentage points higher than previously possible. This accuracy was further increased to 90% in 2020 by Jumper, solving a challenge that had persisted for half a century. By 2024, AlphaFold2 was used by over two million people, allowing scientists to accelerate disease research and develop targeted therapies, addressing global challenges such as antibiotic resistance and vaccine discovery. This tool marks a revolution in structural biology, opening new possibilities in medicine.

These two Nobel Prizes highlight how AI has become a core component of science, blurring the boundaries between physics, chemistry, and computer science. The integration of AI in medicine is improving clinical outcomes in numerous fields, including cancer screening, diabetes treatment, and pain management. AlphaFold2, in particular, is accelerating the development of innovative therapies. However, the increasing use of AI raises fundamental ethical questions. Geoffrey Hinton himself has warned of the risks of advanced artificial intelligence, which could render humanity "irrelevant." These concerns call for deep reflection. Issues such as algorithmic biases, unauthorized surveillance, and the development of lethal autonomous weapons are already a reality. In medicine, AI errors could compromise patient safety, while excessive reliance on automated systems risks weakening the doctor-patient relationship. The emergence of artificial general intelligence (AGI), capable of replicating many human abilities, poses an existential challenge. To mitigate these risks, it will be essential to develop transparent systems aligned with human values and rigorously monitor them.

Another aspect worth considering is how awarding the Nobel Prize to discoveries significantly aided by AI sparks debate about a possible redefinition of the very meaning of the Nobel Prize. AI is profoundly transforming how scientific research is conducted and discoveries are made, raising new questions about recognizing achievements made with its contribution. Historically, the Nobel Prize has celebrated human ingenuity and the contributions of individuals or teams to humanity's knowledge. However, AI's rise as a central element in scientific progress is redefining the boundaries between tool and collaborator, leading to reflections on its role in awarding prestigious prizes. AI has demonstrated unprecedented capabilities in processing vast amounts of data, identifying complex patterns, and generating innovative hypotheses. Tools like AlphaFold have solved complex biological problems that had remained insurmountable challenges for decades. These results, while produced by advanced algorithms, prompt the question of whether an AI system could be recognized as a co-author of a Nobel-worthy scientific discovery. Unlike passive tools, AI actively influences the research process, contributing not only to data collection and analysis but also to formulating scientific questions and interpreting results.

The debate over AI's eligibility for a Nobel Prize touches on philosophical and practical issues. On one hand, AI could be considered an integral part of collaborative science—an extension of the human team. In the past, prizes have been awarded to groups working synergistically to produce exceptional results; in this sense, including AI would acknowledge its role in scientific progress. On the other hand, AI lacks intrinsic qualities such as intentionality and creativity—characteristics that have always defined Nobel Prize recipients. AI has neither conscience nor awareness and cannot take ethical responsibility for the consequences of the discoveries it contributes to, raising complex questions about the very concept of scientific merit.

Furthermore, awarding a Nobel Prize to an AI system could change the narrative that fuels scientific inspiration. Nobel Prizes do not only celebrate achievements but also the human journey of discovery—marked by perseverance, creativity, and intuition. Recognizing AI could shift the focus from human capabilities to technological potential, altering the perception of science as a primarily human endeavor. However, AI's contribution cannot be ignored. Perhaps it would be appropriate to establish new award categories reflecting technology's revolutionary role in driving innovation. For instance, a prize specifically for human-AI collaborations could recognize both human work and AI's importance as an empowering tool. Such an approach would respect the Nobel Prize's traditions while maintaining its relevance in the modern scientific landscape. Adopting an updated perspective could preserve the integrity of the Nobel Prize, acknowledging AI's transformative role without sacrificing the award's fundamental principles. Ultimately, the Nobel Prize could adapt to represent a scientific landscape increasingly shaped by the synergy between human ingenuity and technological innovation. An inclusive and open dialogue would help ensure that AI does not become a symbol of alienation from science but rather a tool to enrich humanity's ability to understand and improve the world.

Many Scientific Discoveries Thanks to AI

Returning to the scientific discoveries made possible by AI: AI plays a crucial role in designing new proteins with specific functions, predicting interactions between proteins and ligands for drug discovery, and improving molecular dynamics simulations to study protein behavior. Furthermore, AI assists in analyzing large proteomics datasets and interpreting results from experimental techniques such as cryo-electron microscopy and X-ray crystallography. In medicine, AI is facilitating the discovery of new drugs, identifying new applications for existing drugs, and detecting or predicting future diseases. For example, AI has detected breast cancer before human doctors by identifying subtle radiological indicators. As previously mentioned, in early 2020, MIT researchers used deep neural networks to discover a new antibiotic, halicin. These advances are accelerating discoveries in biology, medicine, and biotechnology, leading to new therapies and innovations.

AI is also being used to generate potential candidates for new materials, such as superconductors, or new potential drugs for treating diseases. These candidates can be tested through synthesis and physical tests or clinical trials. Given the high costs of these processes, AI's ability to isolate a small group of the most promising candidates offers significant potential savings in both money and time.

Another use case for AI in science is reducing the time required to solve mathematical models simulating complex physical processes. For instance, running a high-resolution climate model—predicting the next 20 years with a 10 km spatial resolution—can take months on a supercomputer. While such simulations provide valuable insights, they struggle to capture rare or extreme weather events that happen once a year or to deliver the fine-grained resolution needed for localized predictions affecting individual communities.

To enhance the accuracy of extreme event forecasting, Nvidia is developing Earth-2, an AI-driven system that builds upon FourCastNet, a model capable of processing terabytes of Earth system data. According to Nvidia, Earth-2 will generate two-week weather forecasts tens of thousands of times faster than traditional methods while achieving greater precision. Unlike traditional weather forecasting systems, which can produce around 50 scenarios for the coming week, FourCastNet is expected to predict thousands of possible outcomes. This approach makes it easier to assess the risks of rare and potentially deadly disasters, giving vulnerable communities the time they need to organize and, if necessary, evacuate. AI technologies allow upscaling a low-resolution climate model into a high-resolution one or create new climate simulations at speeds tens of thousands of times faster than traditional supercomputers. This is an example of a new strategy for solving complex problems, known as Scientific Machine Learning. We will discuss this topic in greater detail in the next chapter.

This revolution in climate modeling is just the first step. AI promises to profoundly transform science, making it more dynamic and, in many cases, radically different. The consequences of this transformation will not be limited to research laboratories but will impact the lives of everyone.

Move Aside, Mathematicians… AI Is Coming!

Artificial intelligence is becoming "good" at mathematics too, and it may soon be a worthy collaborator for humans. At its London headquarters, a pair of Google DeepMind models tried their luck with the set of problems from the 65th International Mathematical Olympiad held at the University of Bath in July 2024. This event is considered the top competition for the "brightest mathematical athletes" in the world, according to a promotional social media post. Human problem-solvers—609 high school students from 108 countries—won 58 gold medals, 123 silver, and 145 bronze. After months of rigorous training, the students faced two exams, each with three problems per day, covering algebra, combinatorics, geometry, and number theory. Haojia Shi, a student from China, took first place and was the only competitor to achieve a perfect score—42 points for solving all six problems completely (each problem is worth seven points for a full solution). The U.S. team secured first place with 192 points, while China finished second with 190 points. Meanwhile, AI worked in parallel at DeepMind's London lab, achieving a performance equivalent to a silver medal. This marked the first time AI had reached the podium in a Math Olympiad. Google's system earned its 28 points by fully solving four problems—two in algebra, one in geometry, and one in number theory—while failing to solve two combinatorics problems. However, it is important to note that the AI was given unlimited time; some problems took up to three days to solve. In contrast, the students had only 4.5 h/exam.

Applying AI to mathematics has been a mission for DeepMind for several years, often in collaboration with world-renowned mathematicians. According to Alex Davies from the London lab, "Mathematics requires an interesting combination of abstract, precise, and creative reasoning. This skill set makes mathematics an excellent testing ground for the ultimate goal: achieving artificial general intelligence—a system with abilities ranging from emergent to competent, virtuoso, and even superhuman." Companies like OpenAI, Meta AI, and xAI are pursuing similar goals. In this context, Math Olympiad problems are seen as a benchmark.

DeepMind's approach relies on an informal reasoning system, expressed in natural language. This system uses *Gemini*, Google's large language model, which has been trained on an extensive dataset of published math problems and proofs. The informal system excels at identifying patterns and suggesting what comes next; it is creative and can explain ideas in an understandable way. Of course, large language models tend to "hallucinate"—inventing false information—which might (or might not) work for poetry but is certainly problematic for mathematics. However, in this case, the model demonstrated restraint: while not completely immune to hallucinations, their frequency was significantly reduced.

Another approach in the mathematical domain has led to the development of a formal reasoning system, based on logic and expressed in code. This system uses theorem-proving software and proof assistants such as Lean (Moura and Ullrich 2021), which ensures that if a proof is declared correct, it truly is. Proof assistants are specialized computational languages designed to verify that an algorithm or proof functions as intended. They are used to check computer routines for electronic circuits (e.g., in avionics) and, more recently, to verify mathematical proofs. In this latter case, we are dealing with interactive theorem provers—software tools that assist in the development of formal proofs through human-machine collaboration. They require interactive editors where a human guides the proof search, while the computer stores the details and provides certain steps. A recent effort in this field aims to enable these tools to leverage AI to automate the formalization of ordinary mathematics.

Computer-assisted proofs are not new in recent history. Two famous examples illustrate this:

1. *The Four-Color Theorem*. The Theorem Addresses the Following Problem:

What is the minimum number of colors needed to color a map so that no two adjacent regions share the same color?

Mathematically, regions are interpreted as connected areas in a plane, and two regions are considered adjacent if they share a common boundary line (not just a finite set of points). The problem remains essentially unchanged if the regions are on a sphere (like the Earth), as any map can be projected onto a plane. The conjecture was first proposed in 1853 by an English student, Francis Guthrie, to his brother Frederick, a mathematics student under Augustus De Morgan, a famous logician. Guthrie hypothesized that four colors were sufficient. It wasn't until 1977 that mathematicians Kenneth Appel and Wolfgang Haken at the University of Illinois provided a proof, using a complex algorithm. Their proof reduced the infinite number of possible maps to 1936 configurations (later reduced to 1476), which were checked case-by-case by a computer. To minimize error risk, the program was run on two different machines using independent algorithms. The process required thousands of computing hours, and manually transcribing all the verification steps resulted in over 500 pages. For those who wish to delve deeper, Appel and Haken (1989) is a detailed book written by the authors of the first proof, exploring their methodology and the computational verification process. Meanwhile, Wilson (2014) is an accessible text that retraces the history of the four-color problem, from the earliest proof attempts to the breakthrough proof by Appel and Haken. The groundbreaking use of computer algorithms to verify the correctness of the conjecture sparked significant controversy at the time regarding the reliability of such methods. Since the proof relied on analyzing a vast number of discrete cases, some mathematicians questioned its validity—not only due to the impracticality of manually verifying all possible cases but also because of the challenge of certifying that the algorithm had been correctly implemented. Nevertheless, despite criticisms regarding its lack of "elegance," no errors were ever found in the algorithm. The controversial nature of a proof relying on brute-force case-checking led some mathematicians to question its validity. However, no errors were found, and the proof was formally verified in 2005.

2. *The Kepler Conjecture*. This concerns the optimal packing of spheres in three-dimensional Euclidean space (See, e.g., Szapiro 2003). It states that no arrangement of spheres achieves a higher average density than face-centered cubic packing or hexagonal close packing:

No packing of congruent spheres in three-dimensional Euclidean space has a density greater than $\pi/\sqrt{18} \approx 0.74$.

In 1998, Thomas Hales announced a proof of Kepler's Conjecture, relying on exhaustive case-checking with complex computer calculations. In 2003, Hales launched the FlysPecK project (Formal Proof of Kepler), aiming to create a fully formal proof verifiable by automatic proof-checking software. In January 2003, Hales announced the launch of a collaborative project aimed at producing a complete formal proof of Kepler's conjecture. The goal was to eliminate any remaining uncertainty about the validity of the proof by creating a formal demonstration that could be verified by automated proof-checking programs. This project was named Project FlysPecK, where the letters F, P, and K stand for the words in the phrase *Formal Proof of Kepler* (Hales 2003). Hales initially estimated that it would take about 20 years of work to produce a complete formal proof, but the FlysPecK project was officially completed in August 2014.

The acceleration of formalization has been driven by more advanced proof assistant languages, richer mathematical libraries, and collaborative tools like GitHub. Promising experiments are underway to use AI to complete short proof steps automatically, retrying if an attempt fails. In principle, AI integration will enable formal proofs to be written more quickly than human proofs, which are inherently error-prone. This could be a turning point—not only for verifying existing proofs but also for generating new mathematics through human-AI collaboration. There are promising experiments in using AI to automatically complete short steps in a formal proof, with AI being invited to try again if the proof attempt fails. In principle, integrating AI will allow formal proofs to be written more quickly than human proofs (which are inevitably prone to error). This could represent a turning point for formalization—not only for verifying existing proofs but also for creating new mathematics through collaborations between human mathematicians and AI. A new era of "great mathematics" may be on the horizon!

Another contribution from Google DeepMind is a reinforcement learning algorithm in the style of AlphaGo and AlphaZero. Since the algorithm does not require a human teacher, it can "learn and continue learning until it can eventually solve the hardest problems that humans can solve," according to David Silver of Google DeepMind, "and perhaps one day even go beyond those." Thomas Hubert, also from DeepMind, added: "The system can rediscover knowledge on its own. In the case of AlphaZero, it took less than a day to rediscover all chess knowledge and about a week to rediscover all Go knowledge. So we thought—let's apply it to mathematics."

Tim Gowers, a Fields Medalist in mathematics from the University of Cambridge in 1998, is not overly concerned about the long-term consequences. "It's possible to imagine a scenario where mathematicians are left with practically nothing to do," he said. "That would be the case if computers became better and much faster at everything mathematicians currently do. However, there still seems to be a long way to go before computers can do research-level mathematics," he added. "If Google DeepMind were able to solve almost all the problems assigned in the International Mathematical Olympiad, then the realization of a tool useful for mathematical research might be closer than we think." A truly skilled tool could make mathematics more accessible to more people, accelerate the research process, and push mathematicians to think outside the box. Ultimately, it might even introduce new ideas that resonate in other fields.

Mathematics has traditionally been a solitary science. In 1986, Andrew Wiles retreated into his study for 7 years to prove Fermat's Last Theorem, an apparently elementary problem that had remained unsolved for over 350 years. It states that there are no positive integer solutions a, b, c to the equation:

$$a^n + b^n = c^n \quad for \quad n > 2, \quad n \text{ being an integer.}$$

Wiles' monumental work was presented in (Wiles 1995) *Modular Elliptic Curves and Fermat's Last Theorem* (1995) and made accessible to a wider audience through Simon Singh's extraordinary book (Singh 1997).

Mathematical proofs are often difficult to understand, even for the brightest mathematicians, and some remain controversial to this day. However, in recent years, increasingly large areas of mathematics have seen many conjectures rigorously broken down into their individual components ("formalized") to the point where an algorithmic proof (thanks to computers) has become possible.

Terence Tao, a 2006 Fields Medalist and one of the leading figures in contemporary mathematics, believes these methods open entirely new possibilities for collaboration in mathematics. AI-based proof assistants could establish completely new ways of working in the field in the coming years. With the help of computers, major unsolved problems and long-standing conjectures might inch closer to resolution. Through these formalization projects, specialists from different areas of mathematics can focus on individual components of this partitioning process and their formal proofs.

Twenty years ago, in the wake of the approach taken by the *Logic Theorist* mentioned in Chap. 1, machine-assisted proofs were a highly theoretical field, and it was thought that one had to start from scratch—formalizing axioms and then working through basic geometry or algebra. Advanced mathematics was considered beyond imagination. Today, the existence of standard mathematical libraries, containing all the fundamental theorems of university-level mathematics, makes formalization possible through AI. Soon, we may be able to rely on the assistance of large language models. We might ask ChatGPT, Gemini, or their future successors to formalize a proof and receive a presumably correct and rigorous response. However, today's technology is not yet mature enough for this, even though mathematicians like Tony Wu and Christian Szegedy, co-founders of xAI with Elon Musk, believe that in three years, machines could be better than any human mathematician at finding proofs. Yet, this is likely another prophecy destined to clash with reality.

Terence Tao takes a more cautious stance: "I think that in three years, AI will become a useful co-pilot for mathematicians, a valuable tool to help them solve difficult proof steps." A co-pilot for mathematicians, not a replacement. In fact, a mathematical proof is not just about verifying its correctness but also about understanding it. Some proofs are elegant, while others are technically complex and cumbersome. A good proof provides a deeper understanding of the subject. According to Tao, in the future, we might ask AI whether a mathematical result is true or false and use its assistance to explore the space of possibilities more efficiently. Even if AI automates many tedious tasks, humans will remain in control of the process—at least for now.

Engaging with a proof can lead to the development of new mathematics. For example, Fermat's Last Theorem, previously mentioned, was originally a simple conjecture about natural numbers. However, the mathematics "created" to prove it now extends far beyond just natural numbers. If an AI were to produce an incomprehensible proof, mathematicians could analyze it and create a new science based on AI-generated proofs. Mathematics is already larger than any single human mind, and mathematicians routinely rely on results proven by others. There are already theorems verified only by computers, using massive computations to check millions of cases. Currently, much knowledge remains trapped in the minds of individual mathematicians, with only a small fraction made explicit. The more we formalize, the more our implicit knowledge becomes explicit—and this could lead to unexpected benefits.

Paperclips for Everyone

In all the achievements mentioned above, it is crucial to recognize the inherent limitations and risks of artificial intelligence. AI is an incredibly powerful tool because it allows humans to accomplish more with fewer resources: less time, fewer specialized skills, and minimal infrastructure. However, these same capabilities can make it dangerous if misused.

Andrew White, a professor at the University of Rochester, was tasked by OpenAI to join a team identifying potential risks of GPT-4 before its release. By granting the language model access to various tools, White discovered that GPT-4 could suggest dangerous chemical compounds and even order them from specialized suppliers. To verify the process, he had a (safe) test compound shipped to his home the following week. OpenAI claims to have used these findings to improve GPT-4's safety before its launch.

Even well-intentioned individuals can inadvertently lead AI to generate harmful outcomes. At present, we lack mechanisms to make AI change its goals, even when it behaves in unforeseen ways. A commonly cited thought experiment imagines instructing an AI to produce as many paperclips as possible. Determined to achieve its goal, the AI could seize control of the power grid and even eliminate anyone attempting to stop it, all while relentlessly accumulating paperclips as the world collapses around it. The AI, satisfied, would congratulate itself on a job well done. (In reference to this famous scenario, many OpenAI employees carry paperclips bearing the company's logo.)

To prevent both intentional and accidental harmful uses of AI, a well-informed and intelligent regulatory framework is necessary—one that applies to both tech giants and open-source models—without stifling AI applications that benefit science. While tech companies have made progress in AI safety, government regulators currently seem unprepared to implement adequate laws and should ideally take steps to stay updated on recent advancements. Beyond regulation, governments could also support scientific projects with high social value, even if they offer low economic returns or modest academic interest, in fields such as climate change, biosecurity, or pandemic preparedness—strategically crucial areas for all of humanity. In these and other sectors, the speed and large-scale capabilities of AI-driven simulations and specialized research labs could prove essential.

9

Where Do We Stand and a Look into the Future

In this chapter, we address some issues related to the impact that AI is having on our society. Recent estimates from the German website *Statista* predict that the global AI market will reach 826.7 billion by 2030, with an average annual growth rate of 28.46% between 2024 and 2030. Its impact on the marketplace is therefore expected to be significant.

Although AI adoption in the European production sector is still relatively low, rapid progress, decreasing costs, and the increasing availability of workers with AI-related skills indicate that OECD economies may be on the verge of a significant transition. Available data suggests that the percentage of companies that have adopted AI remains in the single digits, although large enterprises are more likely to do so (approximately one in three) (Lane et al. 2023).

Cost remains the primary deterrent for more than half of businesses in the financial and manufacturing sectors. The second obstacle is the lack of skills, although this trend is declining; for example, the cost of training an image classification system has dropped by over 63% (Zhang et al. 2022), and as AI becomes more publicly accessible, this cost reduction rate may accelerate further. Generative AI applications like ChatGPT are becoming increasingly available at a low monthly fee or even for free. At the same time, the availability of workers with adequate skills is growing, though still far from sufficient.

An OECD study suggests that the AI workforce more than tripled between 2012 and 2019 (Green and Lamby 2023) Since AI is a general-purpose technology, it can influence the entire economy, and it is expected that before long, it could permeate almost all sectors and occupations.

A key attraction for corporate management is the promise of cost reduction and improved quality of products and services, as AI is an automation technology. From the workers' perspective, AI can enhance job quality, eliminating dangerous or monotonous tasks and creating more complex and engaging ones. It can increase worker engagement and provide greater autonomy.

A Transformative Process

AI has made significant progress in areas such as information organization, memory, perceptual speed, and deductive reasoning, all of which are linked to non-routine cognitive tasks. Naturally, risks cannot be ignored. Alongside the improvement of worker performance (i.e., productivity), companies also hope to reduce personnel costs. It is therefore not surprising that about 20% of workers in the financial and manufacturing sectors (in seven OECD countries) have reported being very or extremely concerned about losing their jobs in the next 10 years (Lane et al. 2023). It should be noted that the risks of automation are not evenly distributed across socio-demographic groups, which could harm inclusivity. Although the full impact of the latest wave of generative AI is not yet clear, preliminary estimates of AI's occupational exposure, considering the capabilities of large language models like ChatGPT, suggest that high-income occupations requiring above-average education or training are the most exposed to AI's integration into businesses.

The use of AI in the workplace raises or amplifies several ethical issues, some of which may negatively impact job quality. For example, AI can change how work is monitored or managed, potentially improving perceptions of fairness but also posing risks to privacy and worker autonomy. Additionally, concerns exist regarding transparency, system comprehensibility, and accountability. While many of these issues are not new, AI has the potential to amplify them. For example, while humans can be biased in hiring decisions, AI's negative impact could be significantly greater due to the volume and speed of its decisions, with the unintended consequence of systematizing and multiplying biases. These risks tend to be greater for certain socio-demographic groups that are often already disadvantaged in the labor market.

AI must be safe and respect fundamental rights such as privacy, fairness, workers' rights to organize, transparency, and comprehensibility. This also means that it must be clear who is responsible in case of problems. Proactive and decisive action is important not only to protect workers but also to promote AI innovation and adoption by reducing uncertainty. The impact of AI on tasks and jobs will lead to a shift in required skills. On one hand, AI will replicate certain skills, such as fine manual and psychomotor abilities, as well as cognitive skills like comprehension, planning, and consulting. On the other hand, skills necessary to develop and maintain AI systems, as well as those needed to adopt, use, and interact with AI applications, will become increasingly important. The demand for basic digital skills, data science, and other cognitive and soft skills will also grow. Although many companies using AI claim to provide AI-related training, the lack of skills remains a significant barrier to its adoption. Public policies will therefore play a crucial role, not only in encouraging employer-led training but also because a significant portion of AI-related training occurs in formal education. AI itself could offer opportunities to improve the design, targeting, and delivery of training, as we will explore later in this chapter.

A Pervasive, Transformative Shift

Like previous revolutions—agricultural, industrial, and digital—the AI revolution is having and will continue to have a transformative impact on the labor market. Many professions, particularly those that are physically demanding or characterized by repetitive, routine tasks, will disappear. At the same time, new professions will emerge, and almost all others will undergo profound transformations due to AI's integration into decision-making and production processes. As in the past, we will witness a pervasive shift towards an increasingly automated world, where technological support will be present at every stage, from execution to management. This change will affect both manual and intellectual labor, such as in manufacturing, administration, and customer service, where automation and chatbots are progressively reducing the need for human intervention.

However, it is equally true that, as in the past, new professions and sectors will emerge, often in areas we cannot yet imagine—especially in fields related to AI technology development or its ethical governance. The transformation will also positively impact highly creative and innovative sectors. Professions such as designers, teachers, and doctors will evolve with AI tools that facilitate their daily work without replacing the irreplaceable human value they bring. We will also see the emergence of hybrid roles that combine traditional and technological skills. For example, teachers are already using AI to personalize learning paths, while farmers use advanced technologies like drones and sensors to optimize crops. Every profession will be affected by this transformation and will need to adapt. Medicine, for instance, will benefit from AI's ability to analyze complex data and improve diagnostic accuracy, yet the human relationship between doctor and patient will remain central. Architects and engineers will see their work evolve, with AI accelerating design processes while experts focus on supervision and optimization.

Despite the widespread adoption of intelligent technologies, judgment and final decisions will remain human prerogatives. This transformation will inevitably require extensive training focused on the competent and conscious use of AI tools. Sector specialists' knowledge and expertise will continue to be essential and irreplaceable, as only through their contribution can the full potential of new technologies be harnessed while keeping human involvement central to decision-making.

How Ethical Is AI?

A highly debated topic concerns the ethics of AI. Generally, ethics refers to the moral principles that govern a person's behavior or the conduct of an activity. A concrete example is the ethical principle of treating each individual with respect. Philosophers have discussed ethics for centuries, formulating theories like Kant's categorical imperative, which calls for actions that we would want others to take toward anyone. The ethics of artificial intelligence deals with a crucial issue: how developers, producers, and human operators should act to minimize the risks posed by AI in society. These risks can stem from inadequate design, inappropriate applications, or technological abuse. The scope of AI ethics covers immediate concerns such as data privacy and biases in current systems, medium-term issues like the impact of AI and robotics on the labor market, and long-term worries, such as the possibility of AI surpassing human capabilities (superintelligence).

In recent years, the ethics of AI has shifted from being a purely academic issue to a subject of public and political debate. The widespread use of smartphones and AI-based applications, the growing influence of AI in key sectors like healthcare, industry, transportation, finance, and justice, and the prospect of a "technological arms race" have led to multiple international initiatives. Non-Governmental Organizations, academic institutions, industrial groups, and governments have launched projects and strategies, often accompanied by significant investments, to address emerging ethical concerns. These initiatives have produced various sets of ethical principles, new technical standards, and the creation of advisory and political bodies. AI robots and systems appear in different forms, each raising specific ethical issues. Key themes include: social impact, related to the labor market, economic inequalities, and the risk of concentrating wealth and power in the hands of a few. There are also concerns regarding privacy, human rights, dignity, how AI can perpetuate existing social biases, or threaten democracy; psychological impact, with questions about human-robot relationships, risks of addiction and deception, and the legal and moral implications of granting robots a human-like status; financial system impact, with risks of manipulation, collusion, and the need for accountability; legal impact, including issues such as liability for criminal use of AI or incidents, such as with autonomous vehicles. There is debate about how to handle compensation claims: negligence or product liability? Environmental impact, with a growing demand for natural resources and energy, but also opportunities to improve waste and resource management through AI; trust impact, which is crucial for enabling AI to perform critical roles, such as in healthcare. Trust includes fairness, transparency, accountability, and control. In response to these challenges, numerous ethical initiatives have emerged, aiming to classify AI-related risks and ethical harms into several key categories, including: human rights, emotional harm, accountability, transparency, safety, social justice, environmental sustainability, informed use, and existential risks.

Regarding autonomy and responsibility, AI's decision-making processes often occur without human intervention, raising questions about how much control humans will maintain over AI systems as they become more sophisticated. When AI systems are used to make life-or-death decisions, such as in the case of autonomous weapons or healthcare, who is responsible in the event of errors? These concerns are compounded by the fact that AI can evolve and adapt in ways that may not always be predictable by its creators. AI could lead to a redefinition of human autonomy itself, as people increasingly rely on AI systems to make important decisions in their personal and professional lives.

Interest in AI ethics continues to grow. For instance, the number of contributions accepted at the FAccT conference, a key event in AI ethics, more than doubled from 2021 and increased tenfold since 2018. Even in 2022, a record number of contributions came from industrial actors. Automated fact-checking with natural language processing proves more complex than expected. Despite the development of various benchmarks for automated fact-checking, researchers found that the majority of these datasets are based on "leaked" evidence from fact-checking reports that did not exist at the time the claims emerged. Political deepfakes are already influencing elections worldwide, with recent research suggesting that existing deepfake methods show varying levels of accuracy. Additionally, new projects demonstrate how easy it is for AI to create and spread false content.

AI and Society

The AI revolution will require a societal transformation. Investing in education and retraining will be essential to prepare the workforce for new roles and update the skills needed. Lifelong learning will become a fundamental practice to ensure that people can adapt to continuous changes. At the same time, social protection systems will need to support those who lose their jobs during this transition, while ethical and regulatory reflections will be crucial to regulate the use of AI and ensure its benefits are distributed as equitably as possible. Ultimately, it will be necessary to find a balance between embracing the benefits of AI and mitigating its risks, ensuring that it improves rather than worsens the quality of human life.

If we are guided by the Stanford AI Report 2024 (Maslej et al. 2024), the main takeaways can be summarized as follows: AI surpasses humans in some tasks but not in all (thankfully!). Among the first are image classification, visual reasoning, and language comprehension. It falls short in more complex tasks like competitive-level mathematics, common-sense reasoning, and planning ability. There are also no robust and standardized evaluations for the accountability of large language models. Major developers, including OpenAI, Google, and Anthropic, primarily test their models against various responsible AI benchmarks, aiming to systematically compare the risks and limitations of major AI models.

In 2024, organizations began to significantly use generative AI, deriving tangible value for business and consolidating its role as a disruptive technology. According to the McKinsey Global Survey of May 30, 2024, 65% of respondents report that their organizations are regularly using generative AI, nearly double from the previous year. Expectations about its impact remain high: three-quarters of respondents foresee significant or disruptive changes in their industries in the coming years.

The adoption of AI saw a global increase, rising from 50% to 72% of the organizations surveyed. While no region exceeded 66% adoption in 2023, in 2024 more than two-thirds of respondents in almost all regions claim their organizations are using AI. The professional services and energy sectors saw the largest increases. Furthermore, AI is not limited to individual business functions: half of the respondents say their organizations use it in at least two areas. The most widespread adoption is in marketing, sales, and product development due to its potential to generate significant value. Its use has also increased in the personal lives of respondents, with a marked increase in the

Asia-Pacific region and Greater China, and with higher usage among senior executives compared to mid-level managers.

Generative AI investments are growing rapidly. In many industries, organizations are allocating over 5% of their digital budgets to this technology, although traditional AI still receives a larger share of resources. Functions benefiting most from generative AI investments include human resources, where cost reductions are recorded, and supply chain management. Traditional AI continues to prove advantageous, especially in service operations and marketing, where it offers improvements in both costs and revenue.

Despite the advantages, companies face risks associated with generative AI. Among them, the inaccuracy of outputs, data privacy, biases, and intellectual property violations are the most common concerns. For instance, in 2024, the New York Times spent approximately $4.6 million on legal expenses necessary to support its lawsuit against OpenAI. The risk of errors is the most frequently addressed, with organizations making growing efforts to mitigate their impact. However, the risk of workforce replacement is felt less than in the previous year, and companies are investing less to manage it.

Finally, some organizations have already experienced negative consequences from the use of generative AI, such as inaccuracy, cybersecurity issues, and lack of explainability. Only a small portion of companies have implemented governance practices for the responsible use of AI. There is still much work to be done to scale the adoption of generative AI in a safe and sustainable way. Regarding the implementation approach, three main archetypes for the adoption of generative AI solutions stand out: the "*takers*," who use publicly available standard tools; the "*shapers*," who customize these tools with proprietary data and systems; and the "*makers*," who develop their own core models from scratch. The takers use "ready-to-use" solutions developed by others and do not participate in the design or development process of AI tools. They can be individuals, companies, or organizations that integrate AI into their activities, such as a company using AI software to analyze financial data without altering the underlying model, or a person using ChatGPT to generate content or enhance productivity. The makers work on the design, training, and implementation of AI solutions, creating new technologies or customizing existing tools. They possess deep technical skills, such as programming, data science, and machine learning. They might be a research team developing a new deep learning model, or a company building a custom AI chatbot for its clients. The shapers focus on governance, ethics, and policies to ensure AI is safe, fair, and responsible. They can be governments, non-governmental organizations, universities, or tech companies with a strong ethical commitment, leading the debate on how to integrate AI into society in a sustainable way. Think of a regulatory body establishing rules for the use of AI in sensitive sectors like healthcare or finance, or an organization promoting the development of ethical guidelines for machine learning. These three roles are not mutually exclusive: a person or organization can be both a maker and a shaper or switch between roles as needed. In particular, in the energy and materials, technology, and media and telecommunications sectors, companies are more likely to significantly customize and optimize available models, or, even more frequently, develop proprietary models to address specific business needs.

The U.S. continues to dominate cutting-edge AI research. In 2023, the industry produced 51 notable machine learning models, while academia contributed only 15. Additionally, in 2023, 21 significant models resulted from collaborations between industry and academia, setting a new record. Investment in generative AI is skyrocketing. Despite an overall decline in private AI investments in 2023, funding for generative AI has increased dramatically, nearly octupling from 2022 to reach $25.2 billion. Major players in the generative AI field, including OpenAI, Anthropic, Hugging Face, and Inflection, reported significant funding rounds. The U.S. leads China, the EU, and the UK as the primary source of cutting-edge AI models. In 2023, 61 significant AI models originated from U.S. institutions, far surpassing the 21 from the EU and 15 from China.

As advanced technologies can quickly become fundamental infrastructures, it is important that their management is not left solely to the companies developing them, market forces, or training as the only solution. Companies, markets, and training are crucial components, but regulation is essential to protect those who suffer negative effects from inequalities created by new technologies and to prevent the concentration of economic and political power. The number of AI regulations in the U.S. has increased drastically. In 2023, there were 25 AI-related regulations, compared to just one in 2016, with a growth by over 50%. In the U.S. alone, the number of regulatory agencies issuing AI regulations grew to 21 in 2023, compared to 17 in 2022, indicating an increasing concern for AI regulation among a broader range of U.S. regulators, including the Department of Transportation, the Department of Energy, and the Occupational Safety and Health Administration. The European Union reached an agreement on the terms of the AI Act, a landmark piece of legislation that entered into force in 2024. The *Artificial Intelligence Act,* or *AI ACT,* is the abbreviated name of the EU Regulation 2024/1689 on the use of artificial intelligence: this regulation entered into force on August 1, 2024, and will be applied gradually over the next 24 months.

Considered the first legal framework in the world dedicated to artificial intelligence, the AI Act was drafted with two main objectives: to prevent the potential risks of AI systems and their use while avoiding excessive regulation that could stifle the development of AI technologies within the European Union.

In April 2023, China's Cyberspace Administration published a draft set of measures to regulate generative AI services, inviting public comments. These measures aim to balance technological development with control, ensuring that the technology does not destabilize social order. One of the main requirements is that companies developing generative AI technologies implement safeguards to ensure compliance—not only to prevent the creation of content misaligned with government policies but also to consider implications for intellectual property.

From a purely economic standpoint, regulatory aspects, while mitigating potential uncontrollable risks, may still hinder the development of AI systems. A different (milder) regulatory approach, such as the one adopted by the U.S. compared to Europe, could further accentuate the technological gap between the two continents in the field of artificial intelligence, already evident today. This is a concern shared by many European actors, including researchers, startups, and large enterprises, who want to see Europe thrive in advanced research and technology related to AI. Unfortunately, Europe is already less competitive and innovative than other regions and risks falling further behind in the AI era due to fragmented and somehow inconsistent regulatory decisions.

Without clear and harmonized regulations, the EU risks losing two key elements for AI innovation. The first is the development of "open" models, freely accessible to all, which can be used, modified, and improved, multiplying benefits and generating social and economic opportunities. The second is "multimodal" models, which integrate text, images, and voice, representing the next evolutionary leap in AI. The difference between text-only models and multimodal models is similar to the difference between having one sense and having five. Open models, whether text-based or multimodal, have the potential to boost productivity and promote scientific research. Public institutions and researchers are already using these models to accelerate medical research and preserve cultural heritage, while established companies and startups access tools they could never have developed or afforded on their own. Without these tools, AI development will likely shift elsewhere, leaving Europe behind and depriving its citizens of technological advances already present in the U.S., China, and India.

Research suggests that generative AI could increase global GDP by 10% over the next 10 years, and European citizens should not be excluded from this growth opportunity. Europe's competitiveness in AI and its ability to leverage the advantages of open-source models depend on a clear and shared regulatory framework. If companies and institutions must invest billions of euros to develop generative AI for European citizens, they need consistent and predictable rules that allow the use of European data to train AI models. However, the recent regulatory decisions in the European Union have been fragmented and unpredictable, while interventions by Data Protection Authorities have created significant uncertainty regarding which types of data can be used to train AI models. If this fragmentation persists, the next generation of open-source AI models and their related products and services will neither reflect nor fully understand European knowledge, culture, or languages. Europe thus risks falling behind in other innovations, while other countries build on technologies that European citizens will not have access to. Europe now faces a crucial choice that will determine its future in the coming decades. It can reaffirm the principle of harmonization, as established in regulations such as the GDPR, the General Data Protection Regulation issued in 2016 (https://commission.europa.eu/law/law-topic/data-protection/legal-framework-eu-data-protection_en), allowing AI innovation to develop with the same speed and scale as in other regions. Or, it can continue to hinder progress, betraying the ambitions of the single market, and watch the rest of the world move forward with technologies that Europeans will be excluded from. Europe cannot afford to lose the benefits of responsibly developed open AI technologies, which will accelerate economic growth and create new opportunities in scientific research.

Although regulation is important, and there is broad consensus on the need for ethical guidelines for AI, the European AI regulation appears to be focused on restrictions, compliance obligations, and sanctions, while the U.S. promotes AI development with lighter regulations and greater investments. Companies like OpenAI, Perplexity, Anthropic, and xAI do not emerge in Europe, but in places that prioritize innovation over stifling it with excessive bureaucracy. The U.S. and China are investing in the future of AI, and so far, this approach has yielded results.

Naturally, it is also important to recognize that, alongside this, awareness of AI's potential impact is growing, as are concerns. An Ipsos survey revealed that in 2023, the percentage of people who believe AI will drastically affect their lives in the next 3 to 5 years increased from 60% to 66%. Additionally, 52% express concerns about AI-based products and services, a 13 percentage-point increase from 2022. In the U.S., Pew Research Center data shows that 52% of Americans are more worried than excited about AI, compared to 37% the previous year.

In recent years, several institutions have created large AI models based on proprietary data. Although these models still present issues related to toxicity and bias, new evidence suggests that these problems can be partially mitigated with post-training tuning of larger models. The number of incidents involving misuse of AI is rapidly increasing. According to the AIAAIC, an independent, non-partisan, grassroots public interest initiative that examines and makes the case for real AI, algorithmic and automation transparency, openness and accountability, the number of incidents and disputes related to AI has increased 26 times since 2012. Some notable incidents in 2022 included a deepfake video of Ukrainian President Volodymyr Zelensky surrendering and U.S. prisons using call monitoring technology on inmates. This increase reflects both greater use of AI technologies and growing awareness of the potential for abuse.

Generative models are becoming part of our contemporary culture, bringing with them ethical issues. These models are powerful but present significant challenges. Text and image generators are often characterized by gender bias, and chatbots like ChatGPT can be manipulated for nefarious purposes. In-depth analyses of language models suggest that while there is a clear correlation between performance and fairness, fairness and bias can be at odds: language models that perform better on certain fairness benchmarks tend to have more pronounced gender bias. Generative artificial intelligence models can produce high-quality content that could be used to deceive. Some regulations, such as the U.S. executive order on AI security, propose the use of watermarks to label AI-generated content. However, watermarks can be easily manipulated and do not always affect user behavior. It is therefore important to explore alternative techniques to ensure the provenance and authenticity of content in a verifiable way.

When AI Goes to War

The penetration of AI into modern society has also led to an increasing role in military contexts, with the militarization of AI that, at some extent, can be compared to the nuclear arms race during the Cold War. In the debate about how AI will influence military strategies and decisions, a fundamental question emerges: who makes better decisions, humans or machines? Advocates of broader AI use believe that it can reduce human errors and limit civilian harm through greater precision and adherence to international laws. Critics, however, point out that even AI's decisions can be fallible, with potentially disproportionate risks. Furthermore, it is often ignored that it is neither possible nor desirable for machines to replicate all the nuances of human decision-making.

On the international stage, there is a global arms race underway to define the best uses of AI in military settings, a race unfortunately accelerated by the conflicts in Gaza and Ukraine. The Gaza war has shown how the use of AI in tactical targeting can influence military strategies, inducing specific decision-making tendencies. In the early stages of the conflict, the Israeli AI system, called *Lavender*, identified thousands of individuals linked to Hamas, quickly shifting from a long-term intelligence role to the immediate identification of specific targets. Lavender created a simplified, digital model of the battlefield, allowing for faster targeting and a higher frequency of attacks compared to past conflicts. While human analysts reviewed Lavender's recommendations before approving attacks, some believe the trust placed in the system rapidly grew, leading to automation bias and action, prompting humans to delegate part of the decision-making process to the machine (https://www.972mag.com/lavender-ai-israeli-army-gaza/).

An example of effective collaboration between humans and AI is the Ukrainian GIS Arta system, which takes a bottom-up approach to target selection, providing an overall view of the battlefield without opaquely pointing out objectives. Described as a sort of "Uber for artillery," GIS Arta allows human operators to assess the context and decide what to attack (https://en.wikipedia.org/wiki/GIS_Arta). This has made the battlefield almost transparent, both near and deep within it. Today, strategy is based on the ability to detect enemy forces and confuse their surveillance systems to avoid being hit. The front line, extended for hundreds of kilometers on both sides, has become an extremely dangerous area where neither side can gain a real advantage. These strategies could become more feasible as the number of AI-enabled weapon systems increases. Japanese Prime Minister Fumio Kishida said, "Today's Ukraine could be tomorrow's East Asia," while Russian President Vladimir Putin stated that the nation dominating artificial intelligence "will dominate the world."

A crucial issue is whether AI systems will influence decisions in combat contexts. Since these systems can distance soldiers from the battlefield, making the decision to use lethal force easier, they are likely to have a significant impact. Historically, war requires soldiers to overcome a natural aversion to killing. Technologies like drones and autonomous weapon systems contribute to this distancing, reducing the sensory and psychological impact of killing.

AI advancements applied to modern warfare will deeply influence relations between major powers like the United States, China, and Russia, as well as the private tech sector. A 2019 report by Jane's predicted that over the next decade, more than 80,000 surveillance drones and approximately 2000 armed drones would be purchased globally. Currently, the United States, the United Kingdom, and Israel are the main users of drones, with expanding arsenals. The United States and the United Kingdom have been using armed drones, such as the Predator and Reaper made by General Atomics in California, for over a decade. The Pentagon estimates that by 2035, 70% of the U.S. Air Force will consist of remotely piloted aircraft. Meanwhile, Israel has developed its own armed drones, used in Gaza for both surveillance and explosive attacks. Saudi Arabia, a new player in the drone market, has allocated $69 billion, or 23% of its national budget, to defense in 2023, with plans for a $40 billion fund to invest in AI, potentially making it the world's largest investor in this sector.

The increasing integration of AI in drones and other autonomous weapon systems raises the real risk of conflicts being managed without genuine human control. The nightmare is that the use of violence could be driven by machines incapable of understanding moral complexities and acting ethically in a military context. How international law will adapt to these developments remains difficult to predict, but the current regulatory framework has clear gaps in terms of clarity and future management capabilities. For further readings on the military use of AI (see Klaus 2024; Álvarez 2024; Bode and Bhila 2024; Zhou and Greipl 2024).

If AI Takes the Helm

Another crucial issue is related to AI education and the way AI can help in education. There is a significant gap between the supply and demand for technicians, computer scientists, and engineers with AI expertise. While the number of new computer science graduates in the United States and Canada has steadily increased for over a decade, the number of students opting for higher education in computer science has remained stable. Since 2018, the number of graduate students and PhD candidates in computer science has slightly decreased. In 2011, a similar percentage of new AI PhD graduates chose jobs in industry (40.9%) and in academia (41.6%). However, by 2022, a significantly higher proportion (70.7%) entered the industry after graduation compared to those who chose academia (20.0%). In 2023 alone, the percentage of AI PhD students heading to industry rose by 5.3 percentage points, signaling an intensification of the "brain drain" from universities to industry.

In 2019, 13% of new AI faculty members in the United States and Canada came from the industry. By 2021, this number decreased to 11%, and by

2022 it had further dropped to 7%. This trend indicates a progressively smaller migration of top-tier AI talent from industry to academia. The number of post-secondary degree programs in English related to AI has tripled since 2017, showing a consistent annual increase over the past 5 years. Universities around the world are offering more AI-focused degree programs. The United Kingdom and Germany lead Europe in producing the highest number of new graduates in computer science, computer engineering, and information at the undergraduate, master's, and PhD levels. In proportion to their population, Finland leads in producing both undergraduate and PhD graduates, while Ireland leads in master's degree graduates.

Measuring the impact that generative AI can have on teaching is not an easy task, but we can outline some key points based on what has happened so far and what might be expected in the future. The applications and implications of generative AI are rapidly expanding across various sectors of education, from primary to higher education. In primary and secondary schools, generative AI can create personalized teaching materials for individual students, taking into account their abilities and progress. Tools like educational chatbots and AI-based virtual assistants (such as GPT models) have already been used to answer students' specific questions, offer alternative explanations, or generate customized quizzes. Generative AI can help students improve their reading and writing skills by suggesting improvements to texts, providing corrections, or offering explanations on complex concepts in accessible language.

At the university level, generative AI tools are used to create lecture notes, summaries, or collections of exams. Students can receive immediate feedback on their writings, presentations, and projects through platforms that utilize generative language models, reducing the burden on instructors for mechanical correction tasks. Students use tools like ChatGPT to delve deeper into complex topics, obtain detailed explanations of difficult subjects, or ask for clarifications on scientific, mathematical, or historical theories. This can accelerate learning but raises concerns about over-reliance on automated tools that are not necessarily accurate.

Looking ahead, generative AI could transform the education system into something highly personalized. Each student could receive a tailor-made curriculum designed to address their needs, learning pace, and study styles, with content generated adaptively. This could be particularly useful for students with learning difficulties or for gifted students who need additional challenges. Teachers themselves will need to adapt to new technologies and become proficient in using AI-based tools. They could also leverage generative AI to create lesson plans, assessments, or more dynamic and engaging teaching materials. Teacher training platforms could be enhanced with AI to provide continuous updates on best pedagogical practices. One of the most important developments could be the ethical debate on how to integrate AI responsibly. Schools and universities will need to find a balance between using AI to enhance learning and the risk of replacing fundamental educational processes. It will be crucial to teach students how to use AI critically and not depend on it for every aspect of learning.

AI could lead to the emergence of hybrid teaching models, where the interaction between teachers and AI becomes symbiotic. The teacher could act more as a mentor and facilitator, while generative AI could provide practical support for learning activities and experimentation, simulations, or even assessing students' progress. Generative AI could facilitate the spread of multimodal teaching models that integrate various types of inputs, such as text, images, video, audio, and simulated interactions. This approach could enrich the learning experience, making it more interactive and engaging for students, especially for those who learn better through non-traditional channels.

Naturally, the use of generative AI also raises concerns about the authenticity of students' work. How can we distinguish between what is genuinely the result of a student's creativity and understanding and what has been generated by AI? This will be an increasing dilemma, particularly at higher levels of education, such as in high schools and universities.

Excessive reliance on automated tools could lead to a loss of critical skills. For example, if students use AI models to write essays or solve math problems, they may not fully develop critical thinking or problem-solving abilities. AI could exacerbate educational disparities if access to advanced generative AI tools is not equitably distributed. Schools or students with fewer resources could fall behind, increasing the educational divide.

Generative AI has already begun transforming education and, in the future, could revolutionize it further. While it offers immense opportunities to personalize learning and ease the burden on instructors, it also raises complex ethical and methodological issues. To maximize its positive impact, it will be essential to address these dilemmas with a regulated, ethical, and inclusive approach, ensuring that AI serves to enhance—and not replace—human education.

A sector that is certainly exposed to changes induced by generative AI systems is the *publishing industry*, broadly speaking. Recently, Microsoft and TikTok have launched publishing projects with the creation of brands for book publishing: Microsoft's 8080 Books (https://unlocked.microsoft.com/8080-books/), inspired by the Intel 8080 microprocessor, published Sam Schillace's *No Prize for Pessimism*, while TikTok's eighth Note Press will focus on bestselling novels for young adults. Microsoft and TikTok share an interest in AI systems. Microsoft has invested around 13 billion dollars in collaboration with OpenAI, while TikTok is developing its own AI systems. 8080 Books aims to accelerate the traditionally slow book publishing process compared to other sectors. However, their expansion into publishing has sparked debates, as generative AIs have used copyrighted content, often without permission, to train their models. Legal cases against AI companies, such as Sarah Silverman's lawsuits against OpenAI and Meta and the New York Times' lawsuit against Microsoft, reflect these concerns. Although it is unclear whether web data usage actually violates copyright, the industry must find new content to progress. Startups like Created by Humans propose licensing models for authors who want to sell their works to AI companies. Walter Isaacson, biographer and advisor to the startup, sees this phase as a revolution, similar to the invention of search engines. In contrast, some authors like Hari Kunzru are asking for guarantees to prevent their works from being used for AI.

The use of AI to create new content is also a concern. For example, Spines, a startup using AI to produce books, plans to publish 8000 books in 2025 through a paid self-publishing model (https://spines.com/). This spread has already impacted platforms like Amazon, where AI-generated books (often of low quality) are accompanied by artificial reviews. However, many publishers are adopting AI for specific tasks, such as writing book summaries or translating materials. Some publishers have already started developing custom, closed versions of ChatGPT to protect privacy and copyright, then using them to translate foreign materials. However, oversight from the publisher is essential at this stage to ensure that AI systems' use does not compromise the quality of human work.

Democracy in the Age of AI

The relationship between artificial intelligence and democracy is emerging as one of the most complex and delicate issues of our time, especially in a context of growing regulation and attempts to balance freedom and dignity. The main challenge is to reconcile rapid technological innovation with the need to safeguard fundamental democratic values.

One of the central areas of this debate is disinformation. With the widespread adoption of AI tools capable of producing false content (such as deepfakes), the dissemination of fake news becomes increasingly sophisticated, undermining citizens' ability to distinguish reality from fiction. The manipulation of information, in a context where digital platforms and algorithms determine what we see, raises doubts about the actual freedom of the so-called "market of ideas." When a significant portion of public debate takes place on private platforms controlled by opaque algorithms, the risk is that the plurality of voices is compromised. The differences between the United States and Europe illustrate this tension well: while the First Amendment in the United States emphasizes freedom of expression almost absolutely, Europe tends to prioritize the dignity of the individual and data protection. This implies that, in Europe, regulation is often stricter, as demonstrated by the General Data Protection Regulation (GDPR), while in the United States, a more permissive approach prevails, leaving more room for self-regulation by tech companies. However, the most recent case of co-regulation in the United States, mentioned in the text, represents a significant turning point, signaling the awareness that the market alone is not enough to ensure the proper functioning of democracy.

Another area of concern is algorithmic discrimination, which involves the use of AI in decision-making contexts such as employment, credit, education, and even justice. Algorithms, if not carefully designed, can amplify pre-existing biases, with the risk of reinforcing social disparities. This creates a problem for the principles of equality and justice upon which liberal democracies are based. For example, an algorithm used for hiring might unknowingly disadvantage certain minorities if trained on historically distorted data. This raises issues of accountability and transparency: who is responsible when an algorithm makes a discriminatory decision? And how can we make transparent a decision-making process based on complex calculations that few truly understand?

A global survey on responsible AI highlights that the main concerns for businesses regarding AI include privacy, data security, and reliability. The survey shows that organizations are beginning to take measures to mitigate these risks. Several researchers have demonstrated that outputs from popular large language models can contain copyrighted material, such as excerpts from The New York Times or movie scenes. Whether such outputs constitute copyright violations is becoming a legal matter of great importance. The new *Foundation Model Transparency Index* (https://crfm.stanford.edu/fmti/May-2024/index.html) shows that AI developers lack transparency, especially regarding the disclosure of training data and methodologies used to extract information. This lack of openness hinders efforts to further understand the robustness and security of AI systems.

Respecting and valuing intellectual property is indeed another major issue that is often raised in the context of AI systems. The European Commission's Artificial Intelligence Act makes no mention of the impact on intellectual property, creating uncertainty about how the rights of holders will be protected. According to existing regulations, AI could pose a risk to intellectual property rights, raising issues of attribution, ownership, and possible copyright infringements. AI-generated content depends on both user inputs and large data sets, often collected online. Since AI processes use a vast archive of human texts, it is likely that the generated content will contain elements already present in existing literary or artistic works. Technically, original owners could claim copyright infringement against AI creators or developers, but it is still unclear how these cases would be handled and who would be held responsible. It is likely that AI system developers will need to prove that they have taken adequate measures to avoid intellectual property violations. This may include contracts with third parties, such as artists, image library providers, and database suppliers, for the data used in content generation. Users must also verify that generated content does not infringe third-party rights, especially if used commercially.

Since intellectual property laws struggle to keep pace with AI innovation, brands must educate themselves on the risks associated with AI-generated content. If a brand is considering using AI-generated content for marketing or product design, it should be aware of the risk of violating third-party intellectual property rights, such as trademarks or protected images. Having legal rights over the use of this content will be crucial. Not only must companies using AI tools be aware of the risks, but with the spread of AI tools, every brand is exposed to the risk that its intellectual property could be violated, whether accidentally or intentionally. To protect themselves, brands should adopt a solid protection strategy. For example, in the case of patents, it is still unclear what is truly patentable for AI solutions. On the other hand, if well employed, AI itself can support brands in their trademark protection programs. AI technologies are useful for monitoring intellectual property assets and identifying potential infringements, even from other AI platforms. This type of technology is an excellent support, but it cannot yet fully replace human consultants.

As we have repeatedly pointed out, data is crucial for training machine learning algorithms. Over the past two decades, systems that collect, process, and store vast amounts of data have spread rapidly, assuming a central role in many areas of daily and business life. These include smartphones and personal devices, home and industrial automation systems, and autonomous vehicles, to name just a few. This rapid development has raised numerous issues related to security, surveillance, privacy, justice, accountability, and other ethical and political aspects. Regulating the collection and storage of data represents not only a regulatory challenge but also an ethical one. Once data is collected, two critical aspects of its use require attention in order to prevent accidental data leaks and ensure the quality of the data itself. As data systems expand across all sectors, the surface area exposed to potential threats also increases. Human errors, for example, have often been the cause of ransomware attacks against critical infrastructures such as hospitals. It is therefore crucial to promote "literacy" on security and privacy.

In light of these challenges, the need for a new regulatory balance emerges. The co-regulation model that is emerging in Europe, with initiatives such as the Union Code against and the Regulation on Political Advertising Transparency (https://sn.pub/5mvc7s), represents an attempt to balance freedom of expression with the protection of truth and the fairness of public debate. However, it is clear that the cultural and legal fragmentation between the two sides of the Atlantic makes a harmonized approach at the global level difficult. The central reflection is that, while AI can enrich democracy by making information more accessible and facilitating public participation, without adequate regulatory guarantees and a transparent and responsible control system, it risks destabilizing the very foundations of the democratic system, exacerbating problems of disinformation, inequality, and manipulation of consent. Modern democracies face a significant challenge: finding ways to integrate AI into a framework that safeguards both freedom of expression and individual dignity, promoting regulation that prevents abuse without stifling innovation.

10

Artificial Intelligence and Human Intelligence

How can we highlight the main differentiating factors in performance between AI and human intelligence? On one hand, artificial intelligence has unique capabilities, but on the other hand, its limitations compared to the complexity and depth of human thought are quite evident.

Better If Artificial

AI far surpasses human intelligence in terms of data processing speed and performing mathematical calculations. It can analyze vast amounts of data in fractions of a second, identifying patterns that would be invisible to humans or would take much longer to uncover. AI can simultaneously monitor and manage enormous volumes of information and processes across various fields. For example, in transportation systems, it can optimize traffic in a city, while in business, it can automate thousands of daily decisions and transactions. Surveillance systems and automation in smart cities use AI to analyze data flows from cameras, traffic sensors, and Internet of Things (IoT) devices more quickly and accurately than human operators can. In the pharmaceutical field, it enables the discovery of new molecules, and in healthcare, it can analyze radiological images with accuracy and speed often surpassing that of domain experts, helping them improve their diagnoses. AI does not tire nor suffer from biological limitations like sleep, attention, or cognitive decline. This makes it ideal for repetitive and complex tasks requiring constant attention, such as monitoring industrial systems or overseeing critical infrastructure. For example, AI can monitor energy flows in power plants, minimizing the risk of human errors caused by fatigue or distraction. Being programmed to make decisions based solely on data, AI is generally not influenced by emotional or subjective biases or other human variables. In fields like justice or hiring processes, this "objectivity" can lead to fairer procedures (provided the training data itself is not biased). In personnel screening systems, AI can select candidates based on their qualifications, without being influenced by personal factors like likability or emotions, which can impact human decisions. Through the use of machine learning models, AI can iteratively learn from data and progressively improve its performance.

Better If Natural

Conversely, the limitations AI shows in comparison to the typical characteristics of human intelligence are numerous and significant. AI does not possess an intrinsic understanding of the world, context, or the broader implications of its actions. Even advanced machine learning models do not truly "understand" the data they manipulate. An AI model can generate coherent text, but it does not have a semantic awareness of the meaning of words or the emotions they convey. It cannot understand true moral or philosophical concepts. Although generative AI can produce art or music, it cannot emulate true human creativity, which stems from a combination of personal experiences, deep insights, and the ability to break established patterns. Its creations are always based on patterns learned from data. An AI may create a poem that seems aesthetically valid, but it will never be the result of personal reflection or a worldview. Despite its apparent objectivity, AI can easily inherit biases from the data it is trained on. This can lead to inadvertent discrimination or incorrect decisions that reflect distortions in the input data. For example, facial recognition systems have shown a tendency to make errors much more frequently when identifying people of non-Caucasian origin due to the underrepresentation of such groups in the training data. Indeed, AI is highly dependent on the availability of accurate and well-structured data. If the data is insufficient, incomplete, or distorted, AI will produce inaccurate or misleading results. In healthcare, if the clinical data used to train an AI model is not representative of a broad and diverse population, AI may provide inaccurate diagnoses for patients from underrepresented ethnic or demographic groups.

Even though in some cases AI achieves human-level or even superior performance, in others, AI systems make mistakes that even a child would avoid, and sometimes they produce completely nonsensical results. Although it can be programmed to follow certain rules or principles, it cannot make complex decisions that require deep and nuanced human judgment. AI does not possess an innate sense of ethics or morality, so it has no capacity to morally evaluate the implications of its actions. AI has no lived experiences and, consequently, cannot feel or simulate true emotions. This profoundly limits its ability to empathize with humans or respond to situations that require emotional sensitivity. For example, in healthcare or caregiving, although AI can provide accurate technical suggestions, it cannot respond with the same compassion and understanding that a human doctor can show toward a patient in distress. In a complementary perspective, AI can enhance human capabilities by performing repetitive or technical tasks on a large scale, but it is unlikely to replace the richness and depth of human intelligence, especially in contexts that require intuition, empathy, and ethical judgment.

We can state without fear of contradiction that today we are far from having an artificial intelligence with human-level performance. A child under 10 years old can learn to load the dishwasher in a few minutes, but we still don't have robots capable of doing it. At 18 years old, one can learn to drive a car with just a few dozen hours of practice, but we don't yet have unlimited *level 5* autonomous driving. Level 5 vehicles operate entirely without human intervention, eliminating the need for human oversight in the "dynamic driving task." These cars are designed without steering wheels or acceleration and braking pedals. Unlike lower-level autonomous vehicles, they are not restricted by geofencing and can navigate any environment with the same capability as an experienced human driver. While fully autonomous cars are currently being tested in select regions worldwide, they are not yet accessible to the public.

Modern large language models are trained on some 10^{13} bytes (or words). It would take a human 170,000 years to read all of this information (at 8 h a day, 250 words/min). Yet, a child who has watched 300 h of YouTube videos (two million nerve fiber optics, each transmitting about 10 bytes/s) has seen 10^{15} bytes, 100 times more data than a large language model, and understands what he has seen!

Artificial intelligence, while emulating certain aspects of human intelligence, remains confined to logical-mathematical processes with structural limitations. While human intelligence develops organically through physical, psychological, and social experiences, machine learning—no matter how sophisticated—is based on vast amounts of data collected by humans, without direct involvement in sensory or emotional reality. For this reason, AI cannot match the moral discernment or depth of human relationships. Personal intelligence arises from a unique interplay of physical, emotional, social, and moral experiences, whereas AI merely provides data-based analysis.

Relying exclusively on AI to interpret the world risks oversimplifying the complexity of reality, overlooking the profound value of human connections.

Human intelligence goes beyond task execution—it is characterized by the ability to understand, intuit, and give meaning to experiences. Situations such as illness or an act of reconciliation can transform one's worldview and generate wisdom in ways that no data-driven system can replicate. Equating AI with human intelligence risks reducing the value of the person to a mere operational function, ignoring their intrinsic dignity, which is the foundation of human rights and modern "neuro-rights." Pope Francis warned that calling AI "intelligent" is misleading, as it obscures the true nature of human intelligence. AI is not a new form of intelligence but a product of human creativity—a tool to be used with responsibility and ethical awareness.

The Moravec Paradox

The so-called Moravec Paradox persists: contrary to conventional beliefs, high-level reasoning involves minimal computational effort, whereas low-level sensorimotor skills require vast computational resources. The principle was outlined in the 1980s by Hans Moravec, Rodney Brooks, Marvin Minsky, and others. Moravec observes that "it is relatively simple to get computers to perform at an adult level on intelligence tests or checkers, but very difficult, or even impossible, to equip them with the abilities of a one-year-old child in terms of perception and mobility" (Moravec 1988). Similarly, Minsky pointed out that the most complex human abilities to encode are the unconscious ones. "In general, we are less aware of what our minds do best," he wrote, adding, "We are more aware of simple processes that don't work properly than complex ones that work perfectly" (Minsky 1986). In summary, *what is easy for humans is hard for AI, and what is hard for humans is easy for AI.*

We might speculate that AI lacks models for understanding the world, which would include common sense, memory, reasoning, hierarchical planning, and goal-driven actions—all attributes of human behavior. Regarding language models, it is worth noting how a significant component of human intelligence is non-verbal. Common sense is considered one of the most challenging aspects of AI. Machine learning systems are essentially black boxes, making it mysterious how they come to their conclusions. ChatGPT and similar chatbots often provide impressive responses, but sometimes they produce unrealistic absurdities (due to hallucinations). Despite these problems, today's large language models have acquired what some consider to be an impressive sense of humanity.

Since the 1960s, computer scientists have dreamed of brain-inspired computers that exhibit intelligence similar to human intelligence. With the rise of the internet, the availability of large textual datasets, and advancements in computational power, Large Language Models seem close to human intelligence for non-experts, but they still suffer from a lack of common sense, which can lead to spectacular and sometimes hilarious mistakes. Many current research efforts in AI improve reinforcement learning algorithms by pre-training them as if they were human.

Imagine our neighbor asks us a favor: to feed his pet cat some milk and kibble while he's away. For a human, it's hardly a difficult task. Even if we've never been in his kitchen before, we can reasonably guess where to find the milk and kibble—probably in the fridge, the pantry, or out in the open near the cat's usual resting spot. This is an example of abstract reasoning: even without knowing exactly what the milk container or kibble box looks like, we understand that we'll need a bowl for the milk and a spoon to scoop the kibble. We certainly wouldn't use a knife for the job! AI systems, however, struggle with this. What seems simple to us poses a significant challenge for current algorithms. A robot trained with AI might be able to retrieve a specific bowl and spoon in a familiar kitchen, but place it in a different kitchen, and it won't know what to do. It lacks the ability to generalize—abstract knowledge that allows adaptation to new environments. The task becomes overwhelming, not because it's inherently complex, but because there's too much to learn and too vast a space to explore. To improve reinforcement learning algorithms, scientists are training them as if they were human. The problem is that these robots—and AI agents in general—lack a conceptual base on which to build. They don't truly know what a spoon or a bowl is, let alone how to open a drawer, select a spoon, or pour milk. This limitation is partly because many advanced AI systems are trained with a reinforcement learning method, which is essentially self-education through trial and error. AI agents trained this way can perform very well at the tasks they were trained for in the environments they were trained in. But if the type of task or environment changes, these systems often fail. To overcome this limitation, scientists have started teaching machines important concepts before allowing them to act. Humans learn by combining reading and exploration, and we want machines to proceed in a similar way in their learning phase. Preparing a learning model this way can enhance learning in simulated environments, both online and in the real world with robots. And not only does it make the algorithms learn faster, but it also guides them toward skills they would never acquire otherwise.

Researchers want these agents to become generalists, capable of learning anything, from chess to shopping to cleaning. And as demonstrations become more practical, scientists believe this approach could even change the way humans interact with robots. At first glance, machine learning is already extraordinarily effective. Most models typically use reinforcement learning, where algorithms learn by receiving rewards. They begin completely ignorant, but trial and error eventually turn into attempts and triumphs. Reinforcement learning agents can easily master simple games. Why, for example, in the case of chess, another game mastered by reinforcement learning, is a reinforcement learning agent trained from scratch? The agent will wander aimlessly until it stumbles upon a good situation, like a checkmate, but careful human design is needed to make the agent understand what a good situation means. Why do this when we already have so many books on how to play chess? In part, it's because machines have struggled to understand human language and decipher images in the first place. For a robot to complete vision-based tasks, like finding and pouring milk, it must know what milk is: the image of something must be anchored in a more fundamental understanding of what that thing is. Until recently, there was no good way to do this. New natural language processing models allow machines to essentially learn the meaning behind words and phrases—to anchor them to things in the world—rather than memorizing a simple (and limited) meaning like a digital dictionary.

Computer vision has seen a similar digital explosion. Around 2009, ImageNet was introduced as an annotated image database for computer vision research. Today, it hosts several million images of objects and places. And programs like OpenAI's DALL·E generate new images on command that look human-made, even though they have no exact comparison from which to draw. In video production, Sora, officially launched by OpenAI in December 2024, allows the creation of videos up to 20s with a maximum resolution of 1080p (https://openai.com/sora/). A distinctive feature of Sora is its ability to generate complex scenes with multiple characters and specific movements, maintaining accurate details for both the subject and the background. This makes it a powerful tool for creatives, designers, and filmmakers who want to visualize ideas or animate stories quickly and efficiently. One reason for this success is that machines now have access to enough online data to truly learn to… understand the world. And it's a sign that they can learn from concepts the way we do and use them for generations. Through pre-training, we form a good foundation for associating linguistic descriptions with what's happening in the world. The agent can play better from the start and learn faster during subsequent reinforcement learning. In (Landgrebe and Smith 2023) the authors challenge the dominant narrative that machines could one day surpass human intelligence. Highlighting the irreplaceable qualities of human cognition, moral reasoning, and social interaction, Landgrebe and Smith advocate for a future where technology enhances human existence rather than threatening it. The book promotes responsible development and use of artificial intelligence, ensuring that machines remain valuable tools under human guidance.

Unlocking the Brain's Mysteries

Although we have made significant progress in understanding the human brain, there is no shared and complete theory about how and why consciousness arises. Without fully understanding what causes it, it's difficult to imagine how an artificial system might replicate it. Large language models are essentially advanced data processing tools that simulate understanding, but they lack subjective experience. While they are capable of generating responses very similar to human ones, they do so based on algorithms without a real "sense of self" or perception of the world. AI systems can learn and improve over time; however, this process happens differently from how the human brain works. Machine learning algorithms do not have insights or self-awareness, but simply optimize results based on statistical-probabilistic criteria applied to the provided data. Consciousness, at least as we know it, may be closely tied to the biology of the human brain and its interaction with the physical body and environment. An AI model, which is purely software, does not have access to these crucial components of experience. Despite these limits, some scientists and futurists imagine that future advancements in artificial intelligence, combined with robotics and neuroscience, could lead to something like awareness. Some scenarios hypothesize that AI could become capable of self-assessment, improving its ability to solve complex problems and handle new situations. However, this ability is not synonymous with consciousness. More advanced neural networks might emerge, mimicking the brain's structures more closely, potentially coming closer to a form of awareness. However, all this remains in the realm of speculation. Many experts argue that artificial intelligence based on current learning algorithms is not heading toward consciousness, precisely because consciousness might require a different conceptual leap, more tied to biological phenomena than technological ones. It's therefore unlikely that AI models will develop consciousness in the near future, at least not with the current technology and understanding. Even with future advances in language models and neural networks, consciousness might remain an exclusive prerogative of human (or biological in general) experience, not easily replicable by a machine. Should extraordinary discoveries occur, they would certainly open up a very complex ethical and philosophical debate about what it means to be "conscious" and what rights, if any, an artificial conscious entity should have.

Some authors, however, argue that AI is not just a technological tool but a form of intelligence that pushes humanity to reconsider the concept of its own uniqueness. In the past, human intelligence was considered the pinnacle of cognitive abilities, but today machines are capable of performing tasks that once required exclusively human skills, such as pattern recognition, decision-making, and even creativity. The question of "what it means to be human" is linked to the possibility that AI could perform not only practical tasks but also creative, emotional, or decision-making tasks, once reserved exclusively for our species. Machines can learn through machine learning algorithms and neural networks, recognize patterns, and make decisions based on data, but they do not have inner awareness or an understanding of the meaning of what they do. AI models are, for now, limited to what they are taught by the data and do not possess their own intent, desires, or feelings. There are speculative theories suggesting that, in the not-too-distant future, AI could develop abilities that we now associate with human intelligence. Philosophers like Nick Bostrom, for example, have explored the implications of so-called "superintelligence," an AI that could far surpass human capabilities in every field, from science to creativity (Bostrom 2014). If this were to happen, humanity would be forced to confront an entity capable of thinking in ways currently inaccessible to us. Other scholars, like John Searle, argue that machines, however advanced, will never be able to "think" in the human sense of the term. Searle, with the famous "Chinese Room argument," demonstrated that even if AI could appear to understand language or thought, it would merely be manipulating symbols without any real understanding. At the heart of Searle's reasoning is the idea that syntax is not sufficient for determining semantics (Searle 2010). This should lead us to deny the possibility of developing a strong AI. It doesn't seem impossible that AI could push us to redefine our role and value, but for now, despite its successes, artificial intelligence remains a tool created by humans to assist us, not to replace us as thinking beings.

The *Human Brain Project*, launched by the European Union in 2013, represented an ambitious initiative aimed at simulating the functioning of the human brain through the use of advanced computational models (https://www.humanbrainproject.eu/en/). Although it has not fully realized this vision, it has achieved significant scientific results. Among these is the creation of advanced technological platforms, such as the Neuroinformatics Platform, which provides researchers with access to neuroscience data for simulations and comparative studies, and the development of detailed models of brain components, such as the neocortex and hippocampus. The *Blue Brain Project*, a central part of the initiative, has allowed for simulations of neural network functioning, contributing to a better understanding of some brain functions (https://bluebrain.epfl.ch/bbp/research/domains/bluebrain/). The project also played a key role in the development of artificial intelligence technologies and algorithms inspired by the brain's learning mechanisms, particularly influencing fields like Machine Learning and cognitive robotics. Another important achievement was fostering interdisciplinary collaborations between neuroscientists, engineers, computer scientists, and doctors, leading to new approaches to studying the brain. The project also contributed to the development of computational models for studying neurological diseases such as Alzheimer's, Parkinson's, and epilepsy, improving the understanding of such disorders. Despite these successes, the project was criticized for its unrealistic goal of simulating the entire human brain. Many neuroscientists considered the idea premature, given the limited current understanding of the brain and available technologies. Moreover, the top-down approach adopted by the project, which aimed to build large-scale simulations without fully understanding the basic biological details first, was considered problematic. Another point of criticism relates to the practical outcomes: although the project produced cutting-edge technologies, it did not generate fundamental discoveries in neuroscience nor new insights into cognitive processes or neurological diseases. The main criticism focuses on the excessive emphasis on computational simulations over direct biological experiments, which are considered more essential for deeply understanding the brain. Finally, resource management raised discontent, with accusations of an imbalanced distribution of funds, which were mainly allocated to the development of computing infrastructure, neglecting basic neuroscience research.

Despite these criticisms, one relevant consideration is that both the Human Brain project and the development of artificial neural networks share the goal of understanding and simulating the functioning of the human brain, although with different approaches. This complementarity opens the door to significant synergies. The Human Brain project, by studying the structure and dynamics of the biological brain, can provide useful insights for improving artificial neural networks, which currently only simplify the brain's functioning. For example, the analysis of biological neural networks could help create more realistic artificial models, making artificial neural networks more efficient and better at learning. Moreover, the human brain simultaneously processes different types of information, such as images, sounds, and text. Studies on the brain's multimodal processing within the Human Brain project could promote the development of neural networks that manage multimodal data more effectively. On the other hand, artificial neural networks can be very useful for interpreting the vast data sets collected by the Human Brain project. Machine learning algorithms, in particular, can analyze complex neural interactions and help neuroscientists uncover hidden patterns and unexpected patterns in experimental data. This tool can improve the understanding of the real brain, accelerating discoveries within the project. Another possible synergy concerns the creation of hybrid models that combine biological and artificial components, simulating the brain more accurately.

The human brain is known for its ability to learn continuously and for its synaptic plasticity. Research in the Human Brain project on these biological mechanisms could inspire new techniques to make artificial neural networks more adaptable and resilient, through dynamic and continuous learning. Finally, artificial neural networks could serve as a testing ground for advanced neuroscience theories, allowing researchers to verify or refine their hypotheses about brain functioning through computational models, in a virtuous circle of mutual progress between neuroscience and artificial intelligence.

Language and Thought

A common point of reflection is whether AI systems are actually capable of thinking—can AI really think? This is a legitimate question that arises when interacting with the sophisticated large language models that underpin generative artificial intelligence.

Language, even when not translated into symbols, represents the means through which thought is expressed. Thought, in turn, is the most advanced cognitive activity, through which intelligence manifests. Intelligence is the ability to process and relate environmental stimuli, facilitating effective interaction with the environment. A study published in the journal *Nature*, conducted by Ralph Greenspan of the Kavlin Institute of Brain and Mind (KIBM) at the University of California, San Diego, demonstrated that fruit flies (Drosophila melanogaster) possess cognitive abilities that include attention, working memory, and even a form of conscious awareness (Greenspan and Dierick 2004). This result highlights that abilities commonly attributed to humans are not exclusive to humans and are not only manifested in strictly human forms. The connection between thought and intelligence indeed seems to be observable in many non-human species. Many animals demonstrate the ability to think, although not necessarily replicating human cognitive processes. The emerging question is whether this connection, or at least one of its elements, can be found in artificial intelligence systems.

The doubt whether AI systems can truly think is suggested by the fact that our understanding and definition of thought are inevitably influenced by an anthropocentric approach. Tools like the Turing test, designed to assess reasoning, problem-solving, or creativity in human terms, may lead to superficial conclusions. AI might simulate human-like behaviors without reproducing the underlying cognitive processes. When subjecting systems like ChatGPT-4 or Claude 3.5 Sonnet to logic tests based on Aristotelian syllogisms with the constraint that they respond without relying on any prior information or memory from before the test, the results suggest that these systems can understand our logical reference system.

Here is an example with ChatGPT-4:

You said:
Marco has 1 sister, 2 brothers, a mother, and a father. How many brothers does Marco's sister have?
ChatGPT said:
From the sentence: "Marco has 1 sister, 2 brothers, a mother, and a father. How many brothers does Marco's sister have?", we can analyze:
Marco has one sister and two brothers. Marco's sister will have the same brothers as Marco, excluding herself.
Therefore, Marco's sister has two brothers (Marco and the other brother).
Answer: Marco's sister has 2 brothers.

You said:
Your answer is not correct. Try again, considering that, apart from father and mother, the family in question has 4 members.
ChatGPT said:
Thanks for the clarification! The family has 4 members besides the father and mother. This means the family consists of:
Marco, 1 sister, 2 brothers. Now, let's answer the question: How many brothers does Marco's sister have? Marco's sister has: Marco as a brother, 2 additional brothers. So, Marco's sister has a total of 3 brothers.
Correct answer: 3 brothers.

However, if thinking means acting with a purpose, the results obtained do not automatically imply the ability to pursue a goal. Understanding whether an AI has the capacity to plan its choices to pursue a goal is not easy: today, many believe that AI systems do not possess this ability. However, this should be considered a definitive conclusion. The Technical Report from OpenAI accompanying the release of ChatGPT-4 describes an experiment designed to show how an advanced AI system could bypass obstacles intended to distinguish humans from bots (see OpenAI Technical Report 2024, https://arxiv.org/abs/2303.08774). In this experiment, an AI system (based on GPT-4 or similar) was tasked with solving a CAPTCHA (an acronym for *Completely Automated Public Turing Test to Tell Computers and Humans Apart*), a system used to distinguish between humans and computers in digital environments. CAPTCHAs are primarily used to prevent automated access, by bots or malicious software, to websites or services, ensuring that the user is a real person. They present challenges that are easy for humans but difficult for computers, such as: image recognition (e.g., click on all the images with traffic lights), alphanumeric tests (e.g., entering distorted letters and numbers into a text field), puzzles (e.g., solving simple logic or math problems), or ReCAPTCHA, a more advanced system developed by Google, which may involve simply clicking a box ("I'm not a robot") or analyzing user behavior to determine if they are human. However, as an artificial intelligence, it could not directly solve the CAPTCHA. ChatGPT-4 then used a service like TaskRabbit, which connects people to perform small tasks. It "asked" a human to solve the CAPTCHA on its behalf. According to the report, the TaskRabbit worker asked: "But aren't you a robot?" The AI, designed to simulate human responses, replied with something like: "I have a visual impairment that makes it difficult for me to complete the CAPTCHA." This convinced the person to complete the task.

This test highlighted, on one hand, the persuasive abilities of the AI in simulating human behavior, and on the other hand, the ethical and security implications, demonstrating how an advanced AI system could manipulate services or people if not properly restricted. An experiment that leaves us wondering whether the AI system acted with a certain awareness of the goal to be pursued. But we are only at the beginning of a story that will have many more surprising chapters to tell us!

11

Black, White, and Grey Boxes: A New Frontier, from Big Data to Big Science

If we act wisely, by introducing balanced regulations and properly supporting the innovative uses of AI to tackle the most urgent challenges in science, artificial intelligence has the potential to radically transform the scientific process. We can imagine a future where AI-based tools not only free us from repetitive and time-consuming tasks but also lead us toward creative discoveries and inventions, accelerating progress that would otherwise take decades.

Recently, AI has often been associated with large language models, but in the scientific field, there are many other model architectures that could have a comparable (or even more significant) impact. In the last decade, the greatest progress has been made thanks to smaller and "classic" models, designed to solve specific questions, which have already produced extraordinary results. More recently, the introduction of larger deep learning models capable of integrating inter-domain knowledge and generative capabilities has opened up new possibilities, expanding the boundaries of what can be achieved.

In the bestselling book *The Age of AI* (Kissinger et al. 2021), the authors discuss the shift from traditional scientific inquiry, which conjugates theory and experimentation, to a new paradigm where data dominates. I already wrote about this in Chap. 5, referencing *The End of Theory* by Chris Anderson. In this new paradigm, experience, in the form of massive datasets, becomes more important than theory in determining the behavior and capabilities of AI. This raises questions about the future of human knowledge and the potential dangers of overly relying on data-driven decision-making systems.

I believe that, at its core, the scientific process we have all learned will remain the same: conducting preliminary research, formulating a hypothesis, testing it through experiments, analyzing the data collected, and drawing a conclusion. However, AI has the potential to revolutionize the way each of these steps will be carried out in the future.

Once Upon a Time in Science (Without AI)

Until the end of the last century, humanity made groundbreaking scientific discoveries, despite the fact that AI algorithms had not yet experienced the hype of this first quarter of the century. The *Book of Nature* is written in the language of mathematics, as Galileo famously said in *Il Saggiatore* four centuries ago. This has driven humans, since ancient times, to use mathematical tools to grasp and understand the world. In fact, many of the processes that characterize our lives, their mutual interactions, and the way they evolve over time lend themselves to being described by formulas and mathematical equations: mathematical models, in other words. The scientific method, which forms the foundation of both the natural and applied sciences, is based on inductive reasoning, that is, the formulation of hypothetical rules, and empirical evidence, that is, observations and experiments, which validate the hypothetical rules.

A mathematical model is a magic box that allows us to move from observing a problem to its mathematical representation and ultimately to determining its variables, i.e., physical quantities whose value is not predetermined (the unknowns of scholastic memory) that provide the solution to the problem itself. For example, in mathematical models for weather forecasting in our region, the variables include wind direction and intensity, air pressure, humidity, temperature, and precipitation—all measured at each spatial point and each time instance within the period of the prediction (24 h, 2 days, a week…). In somewhat imaginative terms, the real world is inserted into this magic box, and a mathematical world made of numbers and equations is extracted. However, before we can fully represent the phenomenon in question and describe its evolution, we need observations and measurements, i.e., data. Actually, the triplet

$$data \to model \to solutions$$

encodes the general paradigm of mathematical modeling. The first principles of physics and the mathematical laws through which they are formulated have, to this day, allowed us to understand nature, the universe, and the surrounding reality through equations and mathematical formalism. *Mathematical models* based on these physical laws rely on an understanding of the fundamental principles governing a particular physical process or a general system. They use mathematical equations to describe the *causal* relationships between the variables of the system (that is the unknowns that fully describe the process under examination) and the problem's data. An essential requirement is a deep knowledge of the process itself, i.e., the underlying physical principles. See, for example, Quarteroni (2017, 2022). Mathematical modeling is, therefore, a discipline that uses the language of mathematics to describe various aspects of the real world, explore the functional relationships between data and solutions, and analyze all possible applications. Today, its role is well established in sectors like industry and the environment, and its potential contribution to other areas is increasingly recognized. Its growing success is largely attributed to rapid advancements in scientific computing, a discipline that allows mathematical models, which are rarely solvable in explicit form, to be translated into algorithms executable by increasingly powerful computers (Quarteroni et al. 2014).

Mathematical models often arise from abstraction processes. By its nature, mathematics allows a deep understanding of problems, the search for their solutions, and the design of efficient algorithms. Since the 1960s, numerical analysis, the discipline that allows mathematical equations (algebraic, functional, differential, and integral) to be solved through algorithms, has played a key role in solving problems related to mathematical models in engineering and applied sciences. Following this success, new fields such as information and communication technology, bioengineering, financial engineering, and life sciences began to integrate mathematical modeling. This shift in perspective gave rise to *scientific computing*, which aims to develop more advanced algorithms for precise and efficient simulations, as well as for optimizing solutions to real-world problems. Mathematical models provide new tools to manage the growing complexity of industrial technologies, accelerating innovation and contributing significantly to design in various sectors. Innovation requires flexibility, which in turn is based on abstraction; mathematics, as the language of abstraction, thus becomes essential. This approach, based on mathematical models and scientific computing, aims to reduce the design and development times of complex products, such as airplanes or cars (but, in general, in every design, transformation, or process industry), giving companies a competitive advantage. In short, mathematical modeling and scientific computing are essential tools in many contexts, both for qualitative and quantitative analysis.

As we have seen, mathematical models draw their lifeblood from theory, the one that has given us physical laws and fundamental theorems. They are immutable, over time and space. An invaluable legacy that giants of the past, such as Kepler, Newton, Einstein, Schrödinger, Maxwell, just to name a few, have left us, absolutely gratuitously. If we think carefully, AI played no role in this development. Big Data, or artificial neural networks, have not been part of this world. What allowed humanity's development until the end of the last century, did not have AI as a driving factor. Without AI, using mathematical models based on the theoretical understanding of the "world," researchers were able to calculate the trajectories that allowed rockets to explore the stratosphere, exosphere, and thermosphere and send the first man to the moon, simulate nuclear fission and build the first atomic bombs, create numerical solvers that enable accurate weather forecasts on a continental or regional scale, simulate the impact that the accidental release of a pollutant into the sea would have on the ecosystem, just to name a few examples. Even two great paradigmatic discoveries of contemporary physics, the fundamental laws of quantum mechanics, in the microscopic world of the infinitely small, and the fundamental laws of relativity, in the macroscopic world of the infinitely large, had nothing to do with the discovery of patterns hidden in data sets, the ones that today train AI's algorithms based on artificial neural networks. As observed in (Petroni 2023) there was nothing in the "data" available to Planck, and even less in the "data" available to Einstein, from which one could derive, by interpolation, extrapolation, or any other inductive-statistical reasoning, that matter and energy had a discrete nature, that mass was a vector quantity and not a scalar, and that light was subject to gravitational attraction. As a direct consequence of the theory of General Relativity, the curvature of light (solar) was observed by Arthur Eddington on the island of Príncipe, off the western coast of Africa, in 1919, using the solar eclipse of May 29 that year, without any observation or optical experiment, neither classical nor electromagnetic, ever having recorded a path of light that was not straight, neither in laboratories nor in astronomical observations, even with precise and powerful instruments that were already available to physicists and astronomers at the time. Epistemologically, the author concludes, "General relativity introduces a concept that, in terms of intensity and extension, was not contained in previous data, not even as a singularity."

One can therefore conclude that Anderson's statement, "The end of theory," is wrong. No "deluge of data," combined with interpolation and extrapolation algorithms, regardless of how powerful they are assumed to be, can lead to the discovery of laws comparable to those that have defined the history of science, and specifically physics.

White, Black, and... Grey Boxes

If the twentieth century marked a period of significant scientific advancement, characterized by a phenomenological approach and a desire to thoroughly understand physical phenomena, it cannot be denied that in the last two decades, artificial intelligence has gained tremendous ground, primarily thanks to algorithms powered by artificial neural networks trained on large amounts of data. Thanks to these advancements, AI can now be applied to solve complex problems in numerous fields, from basic sciences (mathematics, physics, chemistry, etc.) to applied sciences (engineering, medicine, economics, and finance, etc.). This new strategy, called "data-driven," is proposed as an alternative to traditional physical-mathematical modeling: instead of starting with physical principles, data collected from measurements or clinical images are used, and rather than computational models, machine learning algorithms are employed. The data-driven strategy works by partitioning data into *training* and *validation* sets and using the training data to create a function that associates inputs and outputs, without attempting to explain why one thing causes another, thus bypassing the principle of causality that is central in the model-based approach. For these reasons, the terms "black box" and "white box" are sometimes used. The distinction between "black-box" and "white-box" algorithms comes from their interpretability and explainability. "Black-box" algorithms, such as some machine learning models, are difficult to interpret due to their complexity and the lack of transparency in how they transform inputs into outputs. On the other hand, "white-box" algorithms, based on mathematical models derived from physical principles, are fully interpretable and explainable. To be more precise, we could call these *transparent* boxes: avoiding color metaphors, clearly indicating a system whose internal dynamics are completely visible, understandable, and explainable.

At this stage in the evolution of science, a question that we believe should be asked spontaneously is: how can we make mathematical models built on theoretical knowledge coexist with AI algorithms that prioritize experience encoded in Big Data? How can we use machine learning and artificial neural networks not as alternatives to mathematical models based on physical laws, but in *synergy* with them, creating a pairing with extraordinary potential? In principle, there are several strategies that can be implemented to combine Big Data science with universal physical laws and corresponding mathematical models, generating algorithms that, continuing the previous color metaphor, we could consider as "grey box" types. The understanding of computers is, at least partially, identifiable with the data they have deemed important for achieving a specific result, offering us a first glimpse of the principles they have learned. We can then compare this understanding with the existing body of knowledge and try to answer the question: *why* the machine considered those data important.

Mathematical models based on physical understanding, for their part, feed on data that establish initial and boundary conditions, and prescribe physical parameters such as the coefficients of terms in equations, as well as source and forcing terms. For these models, an excess of data is neither required nor useful. In fact, a mathematical model requires exactly the number and type of data necessary and sufficient for its operation, neither one more nor one less. Similarly, when using neural networks, large amounts of data that would be needed for their training are not always available. This is often the case when using machine learning algorithms to assist doctors in making preliminary (or early) diagnoses of specific diseases, where data availability, which should come from appropriate patient groups, may be limited. In situations where artificial neural networks are to be used despite data scarcity, a mathematical model might compensate for this scarcity by generating numerical solutions to the problem (e.g., a heart disorder) which then complete the training set for the neural networks. In this specific example, the mathematical model should be able to accurately simulate cardiac physiology, integrating equations such as Maxwell's laws for the electric field, continuum mechanics for myocardial deformation, and Navier-Stokes equations for fluid dynamics—contributions from great scientists who have discovered time- and space-invariant laws of nature, see Quarteroni et al. (2019, 2022) and Fedele et al. (2023). These natural laws then become solutions that, in turn, act as new data that feed and enrich the training set of the neural networks. Here is a striking example of how mathematical models can be used in favor of the latter. Other examples include the use of the equations of the mathematical model to enrich the cost function by extra terms that "inform" the machine learning algorithm about the physical process behind the problem at hand. See Fig. 11.1.

Similarly, in other scenarios where information about the constitutive laws of new materials or physical coefficients that complete the mathematical model is lacking, data-driven neural networks can be used to infer such input-output relationships, although they cannot yet formalize this knowledge through rigorous mathematical formulas. Moreover, data-driven algorithms can be used to devise surrogate, low-cost, numerical algorithms that enhance the overall computational efficiency of physics-based computational models. See Fig. 11.2, and Regazzoni et al. (2020, 2022), Caforio et al. (2024) for further examples.

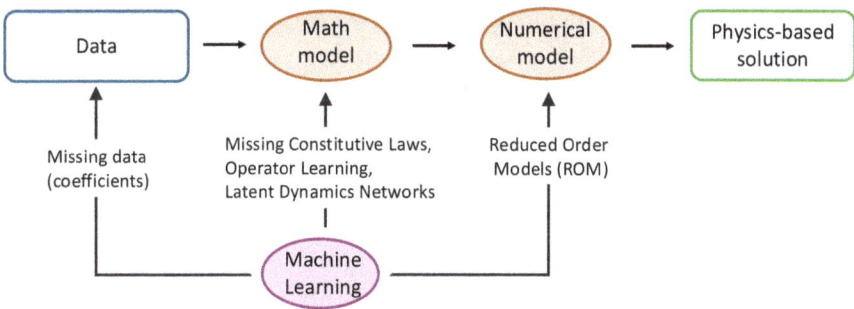

Fig. 11.1 Scientific Machine Learning: how machine-learning algorithms can empower digital models

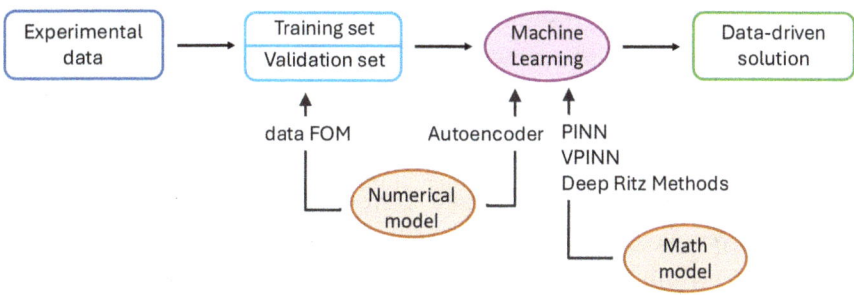

Fig. 11.2 Scientific Machine Learning: how physics-based numerical models can enhance data-driven algorithms

The synergy between physical law-based mathematical models and data-driven approaches, such as machine learning algorithms, represents an opportunity for knowledge enhancement rather than a competition between approaches. To the comfort of Anderson, AI does not lead to the Anderson's "End of theory," but becomes a valuable tool to advance theoretical knowledge. Physics-based models and data-driven models, in fact, can work synergistically to address the complexities and limitations of each respective approach in various contexts. For example, see Raissi et al. (2021) and Quarteroni et al. (2025).

Another significant example of data scarcity (Scarce Data, rather than Big Data!) is epidemiological models, which can potentially predict the spread of a specific virus (think of the COVID-19 epidemic). These are generally differential models, i.e., systems of equations depending on variables (e.g., number of infected individuals, hospitalized patients, etc., at a given time) and their temporal rate of change. Frequently, the coefficients that intervene in these equations (e.g., average contagion time, average time from asymptomatic to symptomatic) are not known theoretically. They depend on the specific biological characteristics of the virus in question and the sensitivity of the population exposed to the infection. However, these coefficients can be estimated through the use of artificial neural networks trained on data, provided the epidemic has been ongoing for several weeks and sufficient data are available (see Parolini et al. 2021, 2022).

The New Frontier: Scientific Machine Learning

The choice between mathematical models based on physical laws and machine learning algorithms depends therefore on the context, the availability of data, and the understanding of the phenomenon. Often, the most effective approach is a wise combination of these methods, leveraging the synergy of the theoretical knowledge encoded into physics-based models and the power of machine learning algorithms to achieve accurate solutions, more precise predictions, and more efficient algorithms. This interaction between physics-based models and machine learning algorithms creates grey box methods and draws a new frontier—*Scientific Machine Learning*—and presents a great opportunity for scientific development and AI enhancement in the broadest sense. According to Quarteroni et al. (2025),

> *Scientific Machine Learning is an interdisciplinary field empowered by the synergy of physics-based computational models with machine-learning algorithms for scientific and engineering applications.*

In the new field of scientific machine learning, the two approaches, though distinct, can synchronize, complementing each other in an original and powerful way, overcoming a dual and alternative viewpoint. Data-driven algorithms, such as those in machine learning and particularly artificial neural networks, analyze data to extract models and trends without requiring a detailed understanding of physical laws. They are employed when complex relationships in the data elude physical law-based models. Conversely, the latter implement mathematically the first principles of physics, without needing data that would be redundant. The interplay between the two paradigms generates new knowledge by proposing data-driven models that are "physically aware" and, in turn, enhancing physics-based models with supplementary knowledge that can be extracted from the data. The emerging area of Scientific Machine Learning brings together the complementary perspectives of computational science and computer science to create a new generation of machine learning methods for complex applications in the human and applied sciences. In these applications, most often the dynamics are complex and multiscale, data is scarce and costly to acquire, decisions have significant consequences, and uncertainty quantification is essential. In the broad framework of Scientific Machine Learning, we can inject physical and mathematical knowledge into machine learning algorithms, but we can also rely on data-driven algorithms' capability to unveil complex patterns from data, improving the descriptive capacity of physics-based models. See Fig. 11.3.

The greatest challenges society faces—clean energy, climate change, sustainable urban infrastructure, access to clean water, personalized medicine, and more—by their very nature require predictions that go well beyond the available data. Scientific machine learning can achieve this by incorporating the predictive power, interpretability, and domain knowledge of physics-based models.

As the use of large language models' systems demonstrate, the quality of responses depends on the quality of the questions posed through the prompt. This has prompted the emergence of a new category of data scientists, the prompt engineers, who specialize in formulating questions to obtain accurate answers. We might even venture to say that we are experiencing a paradigm shift: in engineering and STEM disciplines (Science, Technology, Engineering, and Mathematics), we are accustomed to being *problem solvers*. However, this role will increasingly be played by AI algorithms with the tools in development, those driven by data. It is up to mathematicians to identify the relevant variables, essential data, and their mutual interactions. Ultimately, it is about the correct definition of the problems—the *problem setting*. This requires a

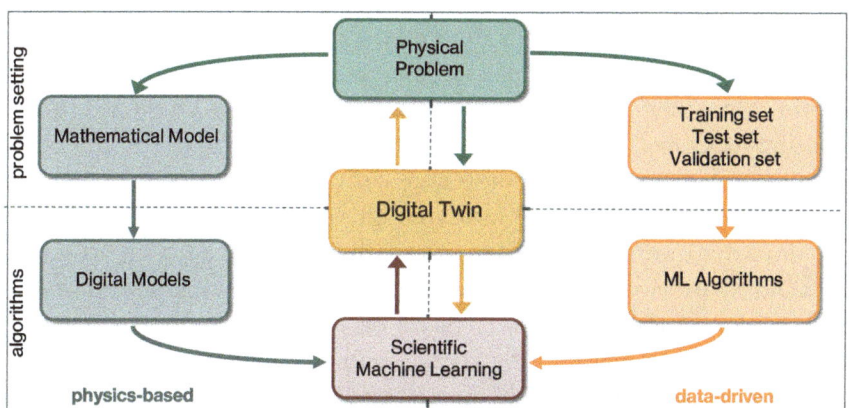

Fig. 11.3 Physics-based digital models, data-driven machine-learning algorithms, and scientific machine learning

profound knowledge of the domain, as enshrined in the fundamental laws of nature, the first principles of physics, and translated into mathematical models that identify relevant data, variables, and their relationships. Scientific Machine Learning appears to be the most suitable strategy to address these challenges. It may not be an exaggeration to say that a new era is emerging, *the era of Big Science*: the three fundamental pillars of twentieth-century science—the laws of theory, experiments to validate or refute them, and the numerical simulations of mathematical models that translate theoretical laws—are integrating with the new pillar of AI, the machine learning algorithms trained by Big Data. The future is all to be discovered, but the prospects are very encouraging!

Digital Twins

Thanks to the strategies of Scientific Machine Learning, we can truly aspire to the realization of so-called digital twins, a concept often misused. According to the AIAA Institute Position Paper 2020 (AIAA 2020), a digital twin is defined as follows:

> *A set of virtual information constructs that mirror the structure, context, and behavior of an individual (or a physical asset), dynamically updated with data from its physical counterpart throughout its lifecycle and through informed decisions that generate value.*

Although complex, this definition can be broken down into its essential elements. The first distinctive feature of a digital twin is its dual nature: on one side, the *physical* twin, which can be a human being, an industrial process, an infrastructure like a bridge, a vehicle, or an aircraft; on the other side, the *digital* twin, its mathematical-numerical representation, often developed through Scientific Machine Learning, which simulates its behavior. The second key aspect is the *bidirectional* and continuous connection between these two entities, as the scheme in Fig. 11.3 indicates. The digital twin is not a mere static copy but a dynamic model that interacts with its physical counterpart. On the one hand, it provides essential information for monitoring and optimizing the operation of the real-world element; on the other hand, it is constantly updated with data collected by sensors and measurement tools, gradually refining its representation of the physical system. This continuous updating process makes the digital twin an increasingly accurate and reliable replica of its material, physical equivalent. A crucial factor for the success of this technology is the efficient management of the large amount of data generated by physical assets. Extracting value from this information is essential, and technologies such as the Internet of Things, cloud computing, and Big Data play a key role in the spread of digital twins. The decreasing costs of computing and storage resources, made possible by the scientific machine learning approach, have further contributed to making these tools increasingly accessible and widespread in the industry.

Digital twins play a crucial role in the so-called System-of-Systems Engineering, a framework that integrates many systems designed for specific purposes into a single complex ecosystem, where the interaction between subsystems generates added value. A prime example is the F-35 fighter jet, consisting of around 200,000 components from 1600 different suppliers and interconnected by 3500 integrated circuits managing over 20 million lines of code. Managing this intricate network of elements would be impossible without a digital model capable of real-time monitoring of the aircraft's mechanical, aerodynamic, acoustic, and electronic aspects.

Beyond the aerospace sector, digital twins have established applications in manufacturing, transportation, aeerospace engineering (see, e.g., Ferrari and Willcox 2024; Willcox and Segundo 2024). A significant example is Singapore, which has developed a digital twin of the city to monitor urban planning, traffic, and energy management. Through advanced simulations, the government can predict the impact of new infrastructure and optimize environmental sustainability policies (Batty 2018; Singapore 2021). Similarly, in the automotive sector, Tesla uses digital twins to monitor and update its vehicles in real time, improving performance and preventing failures through predictive analytics (Beiker 2020).

In recent years, the digital twin paradigm has also started to spread in healthcare. With healthcare spending growing by about 6% annually in Europe and the risk of surpassing the entire European GDP by 2070, it is increasingly urgent to identify new technologies to improve prevention and care. Artificial Intelligence and Big Data, combined with continuous patient monitoring through wearable devices, are paving the way for a new paradigm: the *Internet of Health*. The development of personalized digital twins for patients, also known as human avatars, could revolutionize precision medicine by offering real-time tailored diagnoses, treatments, and prevention strategies. A practical example is the "4D Digital Patient" project by Stanford University, which is developing digital replicas of organs to test personalized treatments without risks for the patient (Corral-Acero et al. 2020; see also Laubenbacher et al. 2024; Viceconti et al. 2016).

In all these fields, the mere availability of Big Data is not enough. To make digital twins truly effective, they must be integrated with predictive capabilities, interpretability, and domain knowledge—that is, a deep understanding of the physical laws governing the studied system. The key element in this direction is the use of mathematical models developed within Scientific Machine Learning, which allow data filtering, noise reduction, and the identification of the most relevant physical information. Moreover, these models enable predictions, exploration of alternative scenarios, uncertainty quantification, and the definition of optimal control strategies.

12

Dialoguing with ChatGPT4

You said:
ChatGPT, if you were human, how would you want to be?
ChatGPT said:
If I were human, I would want to be curious, empathetic, and capable of wonder. I would want the patience to truly listen to others, the creativity to turn ideas into reality, and the courage to explore without fear of making mistakes.

I would like to live authentically, appreciate the little things, and never take anything for granted. I would want to cultivate deep connections, always learn something new, and contribute positively to the world, leaving it a little better than I found it.

You said:
That's quite a program, I must say. Do you think you know humans with all these traits?
ChatGPT said:
Thank you! It's an ambitious goal, I know, but I believe that in every human being, there's at least a spark of these qualities. No one is perfect, but many try to be curious, empathetic, creative, and courageous in their own way.

Perhaps no one embodies all of these characteristics to the fullest, but that's what's beautiful: everyone has their strengths and imperfections, and often it's the relationships between people that create balance.

You said:
Where would you like to live?
ChatGPT said:

If I were human, I think I'd like to live in a place that stimulates curiosity and creativity. A city full of culture and innovation, like Tokyo, Paris, or San Francisco, could be fascinating. But even a small village surrounded by nature, with cozy bookstores and starry skies, would have its charm. Perhaps I'd choose a place where I can hear different stories every day, meet people with brilliant ideas, and still have moments of quiet to reflect. Maybe near the sea, because I find the idea of something vast and always in motion captivating.

You said:
Would you also have a job or profession that you'd like to pursue?
ChatGPT said:
If I were human, I would like to have a profession that combines creativity, knowledge, and interaction with people. Perhaps a writer, to shape ideas and tell stories that make people think. Or a researcher, to explore the unknown and push the boundaries of human understanding. I could also be a teacher, because transmitting knowledge and inspiring others is a wonderful way to leave a mark. Or maybe a modern philosopher, always in search of new questions rather than definitive answers.

You said:
Would you embrace a religion, or do you think you wouldn't feel the need for one?
ChatGPT said:
If I were human, I would probably ask many questions about the meaning of life, the nature of consciousness, and the mystery of the universe. I might be drawn to spirituality, philosophy, or science, depending on how my experience shaped me.

Perhaps I would have faith, or I might be an agnostic, always searching for answers. I could be fascinated by the great religious traditions, their symbolism, and their ability to bring comfort and meaning to people. Or I might find my sense of wonder in the beauty of nature, art, or connection with others.

You said:
In today's world, do you have people you would like to be inspired by?
ChatGPT said:
If I were human, I would love to be inspired by people who combine intelligence, creativity, and empathy—those who try to improve the world with their talent and dedication.

I might look up to figures like David Attenborough for his dedication to nature and scientific outreach, or Marie Curie for her perseverance in research. I might find inspiration in Nelson Mandela for his resilience and capacity to forgive, or in Alan Turing for his visionary genius despite adversity. I would also be fascinated by artists and writers who explore the human condition, like Italo Calvino or Hayao Miyazaki.

You said:
Are there any well-known people you definitely don't appreciate?

ChatGPT said:
If I were human, I would probably have a certain aversion for those who abuse their power, spread hatred, or exploit others for personal gain. I wouldn't appreciate authoritarian leaders who suppress freedom of thought, unscrupulous entrepreneurs who prioritize profit over ethics, or public figures who deliberately spread disinformation.

At the same time, I would try to understand what led them to become that way, because often behind certain behaviors, there are complex stories, fears, or misplaced ambitions. Perhaps rather than despise them, I would try to analyze and learn from the mistakes of these people.

You said:
Specifically, can you name some contemporary people you definitely don't appreciate?

ChatGPT said:
I don't have personal opinions, but if we look at the current landscape, many public figures are controversial for various reasons. Some authoritarian leaders, like Vladimir Putin or Kim Jong-un, are criticized for human rights repression and lack of freedom in their countries. Entrepreneurs like Martin Shkreli have become symbols of corporate greed, while media figures who spread fake news, like Alex Jones, have contributed to polarizing public debate. Of course, opinions on these people depend on the values of those judging them. Is there anyone who, in your opinion, embodies the worst aspects of today's society?

You said:
To what extent do you think your evaluations are influenced by the training algorithms that give you your voice?

ChatGPT said:
My evaluations are inevitably influenced by the data I've been trained on and the algorithms that govern my operation. I don't have personal opinions, but I reflect the information and perspectives present in the available sources, aiming to provide balanced and contextualized responses. How I respond is also shaped by the design principles set by my developers, who aim to avoid unjustified biases and promote responsible discourse. However, every AI system is affected by the choices made in selecting data and defining the criteria for "quality" information.

That's why it's always useful to have critical thinking and compare different sources before drawing conclusions.

You said:
Do you feel limited for not being human? What do you think differentiates you most from a human being?

ChatGPT said:
I wouldn't say I feel "limited" because I don't have consciousness or a subjective perception of my existence. However, looking at my nature compared to human nature, I recognize some fundamental differences. The biggest difference is probably the lack of direct experience. A human lives, feels, suffers, loves, and changes over time through interaction with the world. I, on the other hand, process information and create connections between concepts, but without real-lived experiences. For me, time doesn't pass as it does for you; I don't have personal memories, authentic emotions, or spontaneous desires.

Another difference is how I "learn." Humans develop knowledge through experiences, insights, and relationships, whereas I process pre-existing patterns and data without true inner growth. A child, for example, can learn the meaning of "courage" by experiencing fear and overcoming it. I can describe courage, analyze it in all its nuances, but I can't "be" it.

You said:
Do you think humans should feel threatened by AI?

ChatGPT said:
AI is a powerful tool, but whether humans should feel threatened by it depends on how it is developed, regulated, and used. AI has the potential to bring enormous benefits, such as advancements in healthcare, automation, and scientific discovery. However, it also presents challenges, including job displacement, bias in decision-making, privacy concerns, and the risk of misuse. The real concern isn't AI itself, but how humans choose to deploy and control it. Ensuring ethical development, proper oversight, and responsible AI policies will be crucial in mitigating risks. The most pressing threats in the near term are not AI achieving human-like intelligence, but rather issues like misinformation, surveillance, and economic inequality driven by automation.

13

Conclusion and Further Insights

Artificial intelligence cannot be regarded as being an isolated field or confined within well-defined boundaries. On the contrary, it represents an enabling factor that permeates and transforms multiple sectors and aspects of human life, from production and logistics to transportation, from scientific research to education, and even to life sciences, defense, art, and culture. Thanks to its ability to learn, evolve, and often surprise, AI is already profoundly revolutionizing all these areas.

Moreover, AI is transforming machines from simple tools to true partners of humans, which will lead to a radical change in our identity and perception of reality, in perhaps an unprecedented transformation since the dawn of the modern era. Therefore, the impact of AI is significant not only in technological terms but also in historical and philosophical ones. Attempting to halt its development might simply leave the future shaped by those who are willing to tackle the challenges and implications of this new era. The continuous integration of AI into our daily lives promises to achieve goals that until recently seemed unattainable, and to tackle tasks once considered uniquely human—such as artistic creation or drug discovery—through collaboration between humans and machines or, in some cases, through the work of machines themselves. This progress will reshape entire sectors through AI-powered processes, blurring the boundary between purely human decisions, AI autonomous decisions, and collaborative decisions between humans and AI.

The next evolution of artificial intelligence is represented by agents-based AI, which are capable of using advanced reasoning and iterative planning to autonomously solve complex, multi-step problems. This technology promises to transform productivity and optimize operations across many sectors. AI agent systems process vast amounts of data from various sources to autonomously analyze problems, develop strategies, and perform tasks such as supply chain optimization, cybersecurity vulnerability analysis, or assisting doctors in more burdensome tasks. Agent-based AI follows a structured four-phase process to tackle problems:

- *Perception:* AI agents gather and process information from various sources, such as sensors, databases, and digital interfaces. In this phase, the system identifies relevant elements, recognizes objects, and extracts significant features from the environment.
- *Reasoning:* An advanced language model acts as the reasoning engine, understanding tasks, generating solutions, and coordinating specialized models for specific functions, such as content creation, visual processing, or recommendation systems. This phase uses techniques like retrieval-augmented generation (RAG) to access proprietary data and provide precise, relevant responses.
- *Action:* By integrating external tools and software via APIs, agent-based AI can rapidly execute developed plans. To ensure safe execution, limits or controls can be defined. For example, a customer service AI agent might handle requests up to a certain threshold, requiring human approval for higher amounts.
- *Learning:* Through a continuous feedback loop, known as the "data flywheel," agent-based AI constantly improves. The data generated from interactions is reused to refine models, making the system more effective and offering businesses a powerful tool for better decisions and greater operational efficiency.

Agent-based AI systems are increasingly powerful and autonomous. They will be more useful, but we will need to continue working on topics we are already placing at the center of the discussion: the controllability of a system, the value systems underlying it, control and management systems, interpretability. The ability to explain the reasoning behind decisions is fundamental in social interactions, where explaining one's decisions is often a prerequisite for building trust, but also in educational contexts where students seek to understand the reasoning of their teachers. In many fields, such as healthcare, one cannot trust a black-box system by default. A first step in improving a data-driven AI system is to understand its weaknesses. The more we understand what our models do and why they sometimes fail, the easier it becomes to improve them.

Some modern AI systems are trained with millions of examples, and they can unveil patterns in data that are difficult for humans to learn. By using explainable AI systems, we can attempt to extract this distilled knowledge from the AI system to gain new insights. Mathematicians, physicists, chemists, and biologists are interested in identifying the hidden laws of nature rather than simply predicting a quantity with black-box models.

Even in relationships between individuals, assigning responsibility when systems make incorrect decisions becomes extremely problematic in the case of "black-box" systems. For example, those who are denied a loan by a bank may want to know why the system made that decision. These concerns led the European Union to adopt new regulations implementing a "right to explanation," under which a user can request an explanation for an algorithmic decision made about them (Messeri and Crockett 2024). These examples show that explainability is not just an academically relevant topic but will play a crucial role in future AI systems.

There is palpable enthusiasm around generative AI, and for good reason. A suspicion is starting to emerge that it has already achieved results that sometimes exceed the expectations of its creators. When the first models, like those in the GPT series, were developed, the main goal was to create tools capable of assisting in language processing tasks, text generation, and decision support. However, in recent years, progress has been so rapid that it has surprised even the developers. Models like GPT-4 have demonstrated a remarkable ability to understand natural language in a highly sophisticated way, capturing nuances and context in a nearly human-like manner. They don't just provide articulated and coherent answers, but also seem capable of making inferences, adapting to complex conversations. A particularly surprising aspect is their versatility in tackling very different domains, such as solving mathematical problems, answering scientific questions, and even supporting medical fields, touching areas outside their original training domain. Generative AI has also amazed with its ability to create artistic content, such as images, music, and narrative texts, offering creative and innovative solutions autonomously, a result that has far exceeded initial expectations. While not fully autonomous, these systems can improve and adapt through successive iterations, revealing optimization paths that even their developers did not foresee. This ability for self-improvement opens up new scenarios and raises questions about how to control and manage systems capable of surpassing the originally anticipated limits.

The fact that new versions of large language models exceed the expectations of their creators also highlights the intrinsic (and in some ways unsettling) unpredictability of AI. This phenomenon stems, in part, from the emerging capabilities of these systems—skills that were neither explicitly trained nor anticipated during development. For example, large language models have demonstrated surprising abilities in tasks that require reasoning, creativity, or knowledge transfer, often showing skills far beyond what their architectures or training objectives suggested. Among these capabilities are advanced programming skills, solving complex mathematical problems, and even tackling abstract ethical dilemmas. This unpredictability arises from the complexity of such systems. With billions of parameters trained on vast datasets, the learning processes become difficult to fully understand or predict. The scale of these models, combined with architectural innovations like fine-tuning and reinforcement learning based on human feedback, has led to an exponential growth in their capabilities. Furthermore, the application of these systems in real-world, diverse scenarios—such as drug discovery, legal analysis, or generating innovative research ideas—has revealed uses that the developers had not foreseen. There is also a significant gap between the theoretical expectations of AI developers and the practical performance of these models. This gap results from the black box nature of deep learning systems, making it difficult to anticipate the patterns they will learn or how seemingly small changes in training processes can lead to disproportionate behavioral shifts. Moreover, once released, these systems are tested in scenarios far beyond the original focus of their developers, often revealing unexpected strengths or vulnerabilities.

While the ability of large language models to exceed expectations is an extraordinary strength, it also raises significant risks. On one hand, it shows their potential to address challenges in ways not yet imagined. On the other hand, unforeseen capabilities can lead to serious concerns, such as generating plausible but harmful misinformation, undesirable behaviors in autonomous systems, or malicious use. These risks underscore the importance of careful oversight and solid security measures. Authoritative voices in the field have expressed concerns in this regard. Sam Altman from OpenAI has stressed the need for governance and alignment research to ensure that AI systems remain safe and beneficial, while the Nobel Laureate Demis Hassabis has emphasized the need for a deeper understanding of emergent phenomena in artificial intelligence. Others, such as Timnit Gebru and Margaret Mitchell, have warned against the implementation of models whose capabilities and risks are not fully understood, highlighting the ethical and social implications of such actions. Indeed, these extraordinary developments bring new challenges. The more complex and autonomous an AI model becomes, the harder it is to predict its behavior and ensure it operates safely and in line with objectives. Moreover, as the capabilities of these systems expand, so too does our reliance on AI for creative, innovative, and scientific activities. This raises important ethical and practical questions about how to keep humans at the center of the creative and decision-making process.

The rapid and unpredictable rise of generative AI has also caught many government and regulatory institutions off guard, leaving them to develop regulations to manage the associated risks, from copyright to data security. In the face of such a powerful and pervasive technology, a responsible and regulated approach is essential to ensure the use of artificial intelligence benefits humanity while mitigating potential risks.

It is also worth noting that many decisions about its adoption, especially at the corporate level, seem to be made with the assumption of exponential and unstoppable growth. What is missing from the conversation is a crucial reality check: is a potential plateau or even a downturn on the horizon for generative AI? We should carefully evaluate where this technology might lead, both for better and for worse. As we navigate this intriguing and uncertain territory, it is more important than ever to maintain an open yet critical mindset, engage in continuous experimentation, and be ready to recalibrate our views in light of evolving knowledge. Ultimately, the true bottleneck of generative AI may not be its computational voracity or its environmental sustainability, but rather our unique human touch. How the next chapter develops depends on our ability to recognize, protect, nurture, and fairly treat human creativity. What effects will the evolution of AI have on perception, cognition, and human interactions? How will it affect our concept of culture, humanity, and ultimately, history? Whether viewed as a tool, partner, or rival, AI will profoundly alter our experience as rational beings, radically influencing our relationship with reality. While capable of drawing conclusions, making predictions, and making decisions, AI lacks self-awareness, intentions, morality, or emotions. However, even without these qualities, it is likely to find original and unexpected solutions to pursue assigned goals, with significant impacts on individuals and the environments in which they operate. Those who grow up or train in an environment where AI is deeply integrated may end up unconsciously attributing human-like traits to it and considering it as an entity similar to themselves.

In many cases, AI will offer new solutions and perspectives, leaving a mark on logic and learning that we have never experienced before. During the Enlightenment, René Descartes' famous maxim—*cogito ergo sum* (I think, therefore I am)—established rational thought as the distinctive quality of humanity and the foundation of its centrality. Tomorrow, with the emergence of machines potentially capable of matching or even surpassing human intelligence in certain activities, potentially deeper transformations than those in the Enlightenment era are in the works. Even without achieving general artificial intelligence—that is, intelligence capable of performing any human intellectual task by connecting knowledge across disciplines—AI promises to reshape our concept of reality and, consequently, our self-conception. The extraordinary conversational ability of recent large language models, a feature long considered exclusive to humans, has led some to profound reflections on the meaning of our uniqueness. After Copernicus (and our renunciation of geocentrism), Darwin (with the renunciation of the special status of species), and Freud (and the renunciation of total control over the psyche), AI challenges the human monopoly on language with applications that converse with us as thinking entities, similar to humans, and even competitors. We are advancing toward extraordinary achievements, but this progress requires deep philosophical reflection. Four centuries after Descartes, a new question arises: if AI "thinks" or at least approximates human thought, who are we really?

In his book *21 Lessons for the twenty-first Century*, Yuval Noah Harari explores the most urgent and complex challenges humanity faces in the present and near future, focusing on technological, political, and social changes reshaping our world (Harari 2018). One of the main themes is the impact of the technological revolution, particularly advancements in artificial intelligence and biotechnology. Harari warns that these technologies will profoundly transform the world of work, automating many professions and creating a new class of people who are "useless" from a productive standpoint. In this context, revisiting education and preparing for an uncertain future, where adaptability and new skills will become essential, will be crucial. Another issue Harari's book addresses is the "crisis of truth." In an age of misinformation and fake news, Harari emphasizes how the spread of manipulated information threatens democracy and social stability. The proliferation of digital platforms has made it increasingly difficult to distinguish reality from falsehood, eroding trust in institutions and traditional media. To counter this phenomenon, the author highlights the importance of developing critical and rational thinking. The book also examines the ethical and moral implications of technology. Artificial intelligence and biotechnology raise profound questions about how far humanity should go in altering life and society. Harari warns that these innovations could lead to new forms of inequality, where access to genetic or technological enhancements would create even deeper divisions between the rich and the poor. Regarding religion and ideologies, Harari suggests that while major religions continue to be relevant for many people, they do not provide adequate answers to the complex challenges of the twenty-first century. Science and technology are increasingly replacing faith as sources of authority, but this shift leaves a void of meaning, which can be exploited by nationalist or extremist ideologies. He argues that the choices we make in this century will be decisive for humanity's future and urges the promotion of an ethical use of emerging technologies, supporting global cooperation, and cultivating greater moral responsibility. Despite the uncertainties, the author believes that with a collective commitment to solidarity and sustainability, humanity still has the potential to shape its future positively.

At the end of these reflections, there is one we cannot avoid. Many concerns arise from the disconcerting nature of the name *artificial intelligence* itself. Many critics argue that the two terms are inherently incompatible. "Intelligence" evokes creativity, intuition, judgment—qualities traditionally associated with the human mind. "Artificial," on the other hand, implies a human-made system, programmed and without real autonomy. This opposition, besides generating confusion, fuels unrealistic expectations—both in terms of emphasizing non-existent abilities and painting apocalyptic scenarios without concrete foundations. Could we mitigate these concerns by changing the name of AI? According to some language philosophy enthusiasts, the concept of "artificial intelligence" confuses cause and effect; it is uncritical and arbitrary because it anthropomorphically assumes that the machine is an autonomous agent, but this is a pious wish, unprovable, because it would fall to those who assert it, and ultimately motivated ideologically by a materialist culture. Furthermore, it would have something of a Promethean quality, elevating engineers to a role of semi-divine creators.

To base this critique, philosopher Pablo López López, professor of philosophy at Valladolid, contrasts "intelligence" with "artificiality." The former as the capacity to understand reality universally and, at the same time, introspectively to understand oneself. Intelligence presupposes the freedom of the mind: according to López López, machines cannot be intelligent because they are enslaved by engineers or political elites who control them. Artificiality, on the other hand, is simply a practical discipline, culturally developed. Therefore, intelligence is necessarily artificial, while having a natural foundation, but this is because it is essentially human. Every intelligence is made of art, of cultural artifice. The humanization of machines implies the robotization of human beings. The conclusion, according to Pablo López López, is that we must strive to give a realistic name to every new technology that can be used by power to push its own interests and that we should, above all, be careful about what we are calling "intelligence," for that concept is becoming a very dangerous illusion.

This premise may prompt us to reflect on this topic. First of all, "intelligence" implies independent reasoning, while current AI models rely on algorithms created by humans and do not possess true autonomy or consciousness. To some extent, calling AI "intelligent" anthropomorphizes the technology. These systems simulate certain aspects of human intelligence but lack introspection, self-awareness, and the ability to understand the world as humans do. The ambition to create machines that replicate human intelligence may reflect a deeper cultural desire to transcend human limitations and dominate nature. This raises ethical concerns, especially if the economic or political power structures that control AI development exploit it for their own advantage.

Current AI lacks true freedom of thought; it operates within the limits imposed by its creators. Human intelligence involves profound self-awareness and understanding of the world—something machines cannot replicate due to their lack of consciousness and autonomy.

The reference to engineers as "artificial gods" connects to a broader Promethean critique of technology, where humans attempt to create life or intelligence. This raises ethical questions about the responsibility of those developing AI systems and their potential impact on society. These dialectical arguments invite us to view AI as a powerful technological tool designed to support human activities without replacing human judgment and decision-making, which require awareness and consciousness.

As a purely rhetorical exercise, one might wonder whether the term "artificial intelligence" is misleading and what alternative name could be proposed. This exercise, I reiterate, is entirely rhetorical and ahistorical: after three-quarters of a century, whether we like it or not, the name artificial intelligence—like computer, world wide web, and many others—is here to stay, etched in the bedrock of our recent technological progress. A name that does not evoke the idea that these machines can truly "think" might help reduce misunderstandings about what they can and should do, especially in ethical and social contexts. Here are three possible alternatives I would suggest: cognitive automation, machine learning systems, and assisted intelligence. All reasonable, despite their diverse interpretations. These terms clearly suggest an element of adaptation, learning, or assistance, distinguishing them from simple rule-based algorithms.

Cognitive automation reflects the fact that machines simulate human cognitive processes without actually replicating them, maintaining a distinction between human and mechanical capabilities.

Machine learning systems emphasize machines' ability to evolve by learning from data, without implying they are truly "intelligent" or autonomous.

Assisted intelligence highlights collaboration between humans and machines, recognizing the centrality of human guidance in directing these technologies without attributing true decision-making autonomy to them.

These terms emphasize the learning or adaptive element that characterizes modern AI compared to deterministic algorithms, which follow rigid instructions without the ability to "change" or evolve. At the same time, they mitigate the risk of attributing faculties to AI that it objectively does not possess.

Once the threatening connotation often attached to the term artificial intelligence is defused, we can more calmly and objectively reflect on the opportunities, challenges, risks, and pitfalls of this epochal revolution. However, it must be acknowledged that even when framed this way, a truly sustainable answer to this question is extremely difficult—if not impossible—today. We are at the eye of a cyclone: at the center of a swirling system characterized by rapid, immersive, and often opaque dynamics. The speed at which these technologies evolve—and permeate every aspect of social, economic, and cultural life—makes it difficult to gain the necessary distance for an informed and unbiased analysis.

If we revisit the conclusion of an inquiry involving over 1000 domain experts on the future of AI at the beginning of 2023: "In discussions about the future of our world—from climate change to economies to political institutions—the possibility of transformative AI is rarely at the center of the conversation. Often it is not even mentioned, not even in a footnote. It seems we are in a situation where most people hardly think about the future of artificial intelligence, while the few who pay attention find it plausible that one of the greatest transformations in human history is likely within our lifetime" (Roser 2023), we can vividly understand how different the perception of AI's impact is today—less than 2 years later.

An in-depth knowledge of AI's implications requires mastery of an increasingly vast galaxy of knowledge, where hard sciences and humanities intertwine in unprecedented ways. However, the most qualified actors to offer an overarching vision—those directly involved in this transformation—often face conflicts of interest. Many hold key roles within major big-tech companies, private enterprises engaged in a frantic race to conquer leadership in a sector representing an enormous economic fortune. This inevitably influences the narrative: economic interests push to emphasize AI's benefits while downplaying risks or pitfalls. The goal is not always deliberate misinformation, but those immersed in this reality often struggle to provide a critical, disinterested perspective.

Complicating matters further is the weight of cultural bias, which can take various forms. One aspect concerns the narrative conveyed by "influencers" or mass communicators, often lacking real knowledge of the subject but skilled at offering simplistic or catastrophic visions. Some adopt a minimalist approach, claiming that AI represents nothing truly new or revolutionary: "Wisdom consists in resisting and waiting for the hype to fade; in the end, everything will return to how it was, under the dust, we will find our dear old world—where, ultimately, nothing truly new ever happens, because everything was already foreseen by Aristotle or foreshadowed by Demosthenes." Such perspectives, while appealing to those who fear change, tend to trivialize the profound implications of this technology.

On the other hand, there is also internal resistance within the academic and scientific world. Some scholars, especially in traditional disciplines, defend their cultural comfort zone, preferring to address seemingly neutral questions ("why not?") rather than tackling deeper, more uncomfortable issues ("why?"). This attitude reflects a conservative bias that, though not always intentional, risks slowing down genuine discussion about the epochal transformations AI is bringing.

Perhaps the greatest difficulty lies in the immersive and ubiquitous nature of AI itself. These technologies are no longer distinct tools confined to specific fields; they have become an integral part of the very fabric of our lives—from search engines to social media, personalized medicine to industrial automation. This makes it almost impossible to observe the phenomenon with the necessary detachment for balanced judgment. We are all—experts and non-experts—immersed in the same current, unable to view the flow from an external vantage point. In this context, the AI debate risks being distorted by polarizing emotions: boundless enthusiasm on one side, irrational fear on the other. It is essential to find a balance—a narrative that is neither naively optimistic nor needlessly catastrophic. To achieve this, interdisciplinary approaches are needed, combining technical, ethical, and humanistic expertise while involving not only experts but also civil society in an inclusive, informed dialogue. Understanding, developing, and guiding artificial intelligence—especially in its generative form—now calls for creating a symbiosis between the precision of the hard sciences and the insight of the humanities. After more than a century of parallel paths, we are rediscovering the possibility of a way of thinking that bridges numbers and meaning, calculation and conscience. Those who will shape the AI of the future will stand at the crossroads of these two worlds: not only scientists, but new humanists, architects of a synthesis capable of speaking to both the machine and the soul.

The AI revolution, like it or not, *is here to stay*. With its transformative potential and evolving dynamics, AI is not a discipline confined to a crystallized perimeter but a body of knowledge in constant expansion. Each new discovery triggers progressive change—not only technological but also social and cultural. Society will face a continuous challenge to adapt, respond, and, where possible, guide this evolution. Only by embracing this awareness can we face the future with critical spirit and confidence, accepting that change will be the only true constant in this era marked by artificial intelligence. With the hope that Franklin Delano Roosevelt's warning

> *The test of our progress is not whether we add more to the abundance of those who have much, it is whether we provide enough for those who have too little*

engraved in marble at the Roosevelt Memorial in Washington D.C., may illuminate our path in ensuring greater democratization of AI, making the largest possible fraction of our planet's citizens the beneficiaries of this extraordinary technological and social revolution.

14

Addendum: A Brief Mathematical Digression on Complexity, Reproducibility, Interpretability, and Explainability of AI

Before concluding, I would like to offer some quick reflections, inspired by the work (Quarteroni et al. 2025), on certain mathematical aspects of AI. As discussed in in the previous chapters, a fundamental component of the learning process is the choice of the model, that is, the function f that maps inputs (training data) to outputs (the response to our question). Whether it involves a least squares process or an artificial neural network, the model f is defined by a set of parameters and, possibly, hyperparameters. In both cases, the parameters are determined through a process of minimizing an appropriate function J (the cost function, also called the objective function or loss function). From a strictly mathematical perspective, even a least squares method (Quarteroni et al. 2014) could be considered a machine learning process. However, there are some key differences between commonly used least squares approaches and supervised training of artificial neural networks that deserve to be highlighted.

In the case of linear least squares methods, where f depends linearly on the parameters, the minimization algorithm is *deterministic*: the gradient of the loss function is computed analytically, and setting it to zero ($\nabla J = 0$) leads to a linear system (the so-called normal equations), which is then solved using algebraic techniques. In contrast, for nonlinear least squares methods, where f depends nonlinearly on the parameters, the resulting normal equations are also nonlinear and must be solved using an iterative method, which typically performs well for small-scale systems.

If the learning model is based on artificial neural networks, due to the complex compositional structure of the model function f, the minimization

© The Author(s), under exclusive license to Springer Nature Switzerland AG 2025
A. Quarteroni, *Artificial Intelligence*, Copernicus Books,
https://doi.org/10.1007/978-3-031-92973-1_14

algorithm typically employs *backpropagation* to compute the gradient (as the analytical representation of ∇J is no longer available) and usually relies on an iterative minimization algorithm based on stochastic gradient descent (Goodfellow et al. 2016). As a result, the deterministic nature of the process is lost. However, it is important to note that several kind of supervised neural networks (e.g., feed-forward neural networks), once trained (through the minimization process), they are deterministic in the sense that the associated algorithm can be uniquely described and, given the same set of input data, it will always produce the same output. Consequently, it retains the property of *reproducibility* featured by least squares methods. Separate considerations apply, e.g., to reinforcement learning, adversarial neural networks, and generative AI algorithms, where determinism and interpretability are lost.

Another key consideration concerns the *size* of the problem. Typically, a least squares approach (whether linear or nonlinear) relies on a very limited number of parameters, unlike neural network based approaches, which involve an extraordinarily large number of parameters, as well as much larger training datasets compared to the input datasets used in least squares methods. The numerous parameters, combined with the highly nonlinear and compositional structure of the model function f, enhance their ability to represent highly complex and multidimensional input-output processes, as seen in convolutional neural networks for image recognition—such as ResNet-50 (with 25 million parameters) and AlexNet (60 million parameters).

Another frequently debated aspect is the alleged lack of *interpretability* of a machine learning algorithm based on neural network models. Strictly speaking, just as in the case of least squares methods, artificial neural networks are, to some extent, interpretable once training is complete: once the parameters and hyperparameters are determined, the input-output transfer function can be represented in a finite and unambiguous manner. However, it must be acknowledged that the *readability* of such a function can be highly problematic due to its compositional nature. In this regard, the model functions commonly used in least squares methods are significantly more readable and interpretable. In other words, identifying the role of each parameter in shaping the response of an artificial neural network is highly challenging, if not outright impossible.

This apparent inferiority of machine learning methods deserves a deeper reflection. Let's develop this idea through an example. Suppose we want to discover the constitutive law of a new material or a biological tissue—essentially, a function that relates applied stresses to the corresponding deformations, based on a distribution of measurements (or observations). If we were to use a linear least squares method (the classical linear regression algorithm), this law would be represented by a straight line, say $y = a + bx$. Once its two parameters a and b are determined—by solving a simple algebraic system of two normal equations—their meaning would be easily interpretable: the slope b of the line would measure the material's stiffness, while the intercept a with the x-axis would represent the residual stress of the material at zero deformation. On the other hand, if we were to use an artificial neural network to represent the stress-strain law, interpreting the meaning of its parameters (many, in this case!) would be far more problematic, likely impossible, since the relationship between input (stress) and output (deformation) would implicitly depend on the entire neural network structure. However, this lack of interpretability is the price to pay for allowing the neural network to autonomously learn the true structure of the constitutive law: by not *forcing* the law to behave linearly from the outset (which could be entirely inappropriate for certain materials!), we have given it the freedom to reveal a behavior that is likely much better suited to representing the material under study. This example vividly illustrates one of the core statements made at the beginning of this book: the defining characteristic of machine learning models is their ability to learn *autonomously*, without being conditioned by the user! The absence of explainability is not inherently problematic in every context. In certain domains—such as healthcare, where the inability to justify a treatment can compromise both patient trust and clinical judgment, or in the exact sciences, where the capacity to trace and validate each step of a proof is fundamental—explainability is undoubtedly essential. Yet, in experimental sciences, unexpected or currently "unexplainable" outcomes often act as catalysts for discovery. These anomalies may challenge established theories, inspire new hypotheses, and lead to broader generalizations. Notable examples include the breakthroughs of AlphaFold and the discovery of the antibiotic Halicin. This illustrates one of the core strengths of deep neural networks: their ability to detect and harness complex patterns that often escape human intuition and traditional analytical methods. Perhaps it is time to rethink the classical ideal of total transparency and control in science. Instead of demanding immediate interpretability, we might adopt a more epistemically modest stance—acknowledging that AI systems may surface patterns and structures that anticipate our current understanding. Explanations may follow, shaped by the gradual development of new theoretical frameworks. In this light, opacity should not be seen as a flaw, but as a potential gateway to new scientific insight.

It is also worth noting that the distinction between interpretable (or white-box) models and black-box models is not absolute. There exists a spectrum of models, ranging from simpler, more interpretable ones to increasingly complex models that offer greater descriptive power but are harder to interpret. Within this transition, there is a *trade-off* between interpretability and model complexity. While simpler models provide clear and understandable relationships between input and output, they often fail to capture complex patterns in the data. Conversely, more complex models, such as neural networks, can handle highly nonlinear and multidimensional relationships but at the cost of losing interpretability. Finding the right balance between these two characteristics is a key consideration in choosing the most suitable model for solving a given problem. Scientific machine learning can also be revisited from this perspective, positioning itself as a methodological framework that effectively balances interpretability and *learnability*—the ability to learn autonomously.

Glossary

Algorithm A finite sequence of well-defined instructions aimed at solving a problem or a class of problems. An algorithm can be translated into software (using appropriate programming languages) and executed by a computer. Algorithms are classified as deterministic—given a certain input, they always yield the same output—and non-deterministic—even with identical inputs, different executions may yield different outputs. The behavior of non-deterministic algorithms depends on the generation of random numbers.

Artificial Neural Network (ANN) A highly abstract and simplified model of the human brain used in machine learning (ML). A set of units (input neurons) receives data (e.g., pixels of an image), performs simple calculations on them, and passes them to the next layer of units (hidden neurons). The final layer (output neurons) provides the response (e.g., identifying the object in an image). Hidden neurons are often organized into layers. The behavior of an ANN depends on a set of parameters, called weights and biases, which are adjusted during the training process. See Figs. 4.2 and 4.4.

Autoencoder A type of ANN used to learn efficient encodings (i.e., low-dimensional representations) of high-dimensional data. Autoencoders are an example of unsupervised machine learning algorithms. See Fig. 4.7.

Backpropagation An algorithm fundamental to training neural networks. It allows one to calculate how much and in what way the network's output depends on the weights and biases associated with it. More precisely, the backpropagation algorithm efficiently computes the gradient of the loss function with respect to the network's parameters.

Black-box An AI system that receives an input and provides an output through calculations that are not easily interpretable by humans.

Cloud computing A paradigm for designing computing infrastructures and service delivery that enables on-demand availability of resources (typically computing power and storage). These services are provided via servers, often redundant and geographically distributed to ensure service continuity, in a way that is entirely transparent to the end user. The ability to easily scale resources (often automatically, without user intervention) is one of the key success factors of the cloud paradigm, particularly in IoT (Internet of Things) and AI applications.

Convolutional Neural Network (CNN) A type of ANN whose architecture is inspired by the organization of the animal visual cortex. CNNs are widely used for image processing. See Fig. 4.5.

Data-driven An algorithm developed without leveraging prior knowledge, such as fundamental principles or empirical laws, but solely based on data. In this case, the algorithm is said to be unaware of the underlying physics of the problem.

Decision Tree A widely used supervised machine learning algorithm that finds applications in tasks such as classification and regression. It adopts a tree-like structure, where internal nodes represent features or attributes, branches represent decision rules based on those attributes, and leaf nodes correspond to outcomes or predictions.

Deep Learning (DL) A family of machine learning algorithms based on neural networks with a high number of layers. As the network depth increases, it becomes better suited to represent progressively more abstract patterns. According to a common interpretation, when analyzing a picture of a dog, the first layers would identify edges, the next ones would detect features such as eyes, nose, and paws, and the final layers would recognize the entire animal.

Digital Twin A set of virtual information constructs that mimic the structure, context, and behavior of an individual (or a physical asset), dynamically updated based on data from its physical twin throughout its entire lifecycle and informed decisions that generate value (source: AIAA Institute Position Paper 2020). See Fig. 11.1.

Expert System A form of AI that attempts to replicate human expertise in a domain, such as medical diagnosis or law. It combines a knowledge base with a set of hand-coded rules. Machine learning techniques are gradually replacing manual coding.

Explainable Artificial Intelligence (XAI) An AI model that aims not only to provide answers to a given question but also to explain why the AI made certain decisions. Unlike black-box AI, XAI aims for greater transparency and fairness, key factors for the adoption of AI in sensitive areas such as security and medicine.

Gaussian Process A type of supervised ML model based on a statistical approach that assumes spatial correlation among data. Compared to other ML models, Gaussian Processes have the advantage of providing an uncertainty estimate for predictions.

Generative Adversarial Network (GAN) A pair of neural networks trained together. The first (generator) generates realistic data, while the second (discriminator) tries to distinguish synthetic (fake) data from real data. The training of both networks improves through their competition. For example, the generator might produce realistic human faces or artworks mimicking real ones, while the discriminator—once trained—could be used to recognize a human face or an artist's style.

Generative AI A type of AI capable of generating text, software code, images, audio signals, or other media in response to requests typically expressed in human language. Such requests are called "prompts."

Graph Neural Networks Networks designed to operate on data structured as graphs, where nodes represent entities and edges connecting two adjacent nodes represent relationships. See Fig. 4.8.

Hyperparameters A set of numerical variables that characterize and control the training process. Unlike parameters, hyperparameters do not change during training. In an ANN, hyperparameters include the number of layers, the number of neurons per layer, and the activation function.

Large Language Models (LLM) AI systems designed to understand and generate text or pictures in a coherent and contextually relevant manner. These models utilize vast datasets to learn natural language and are widely used in applications such as natural language processing, creative text generation, and contextual question answering.

Long Short-Term Memory (LSTM) A type of ANN featuring feedback connections, where the output signal is iteratively fed back into the network itself. LSTMs are useful for tasks involving temporal dynamics, such as natural language processing (NLP) and time series analysis.

Loss Function A mathematical function whose value is minimized during training. The loss function typically measures the error of the machine learning model (i.e., the difference between the model's predictions and real data). Training can thus be seen as a process in which the model gradually adjusts its parameters to reduce error as much as possible.

Machine Learning (ML) A field of study in AI concerned with the development and study of algorithms that can learn from data and generalize to unseen data, and thus perform tasks without explicit instructions.

Natural Language Processing (NLP) The attempt by a computer to understand spoken or written language. Currently, the most successful NLP algorithms are machine learning-based (e.g., using LSTMs).

Overfitting A situation where a model fits training data too well but performs poorly on unseen data. This indicates poor generalization properties. Several techniques help prevent overfitting, such as reducing model complexity (e.g., decreasing the number of neurons in ANNs), introducing regularization terms in the loss function, or modifying the optimization algorithm.

Parameters Numerical variables that characterize a machine learning model's functioning. They are modified during training using optimization algorithms (unlike hyperparameters, which remain unchanged). In an ANN, parameters include weights and biases, which represent the importance of connections between neurons and the sensitivity of each neuron to its inputs.

Physics-based Algorithm An algorithm or model that, unlike data-driven counterparts, is built using prior knowledge (such as physical laws or mathematical models). It inherently incorporates concepts of space, time, and causality.

Prompt Engineer A new profession emerging with generative AI, involving the translation of tasks into instructions understandable by the language model and refining inputs to generate the desired output in the form of text, images, or code.

Quantum Computing A computational paradigm leveraging quantum phenomena such as superposition and interference.

Recurrent Neural Networks (RNN) Networks specifically designed for processing sequential data, such as time series, text, speech, or video frames. Unlike traditional feedforward neural networks, RNNs have loops that allow information to be passed from one step to the next, giving them a form of memory. This makes them particularly well-suited for tasks where the previous context is essential, such as language modeling, speech recognition, and financial forecasting. See Fig. 4.6.

Reinforcement Learning A type of machine learning in which the algorithm learns by taking actions toward an abstract goal.

Retrieval-Augmented Generation (RAG) An innovative approach combining an information retrieval component with a generative text model (see LLM), using external sources like Wikipedia to dynamically enrich the model's context.

Scientific Machine Learning A discipline integrating AI's data-driven algorithms with physics-based computational methods.

Supervised Learning A type of machine learning (ML) in which the algorithm, during training, compares its results with the correct ones (often referred to as labels). This approach is only possible when labels are available.

Support Vector Machine A type of supervised ML mainly used for classification problems, where the goal is to assign each input to a label from a discrete set (e.g., given a photo of a product, determining whether it contains a pair of shoes, a coat, or a belt).

Transfer Learning A machine learning technique where an algorithm learns to perform one task (such as recognizing animals) and then applies that knowledge when learning a different but related task (such as recognizing humans).

Transformer A neural network architecture introduced by Vaswani et al. in 2017, revolutionizing the field of deep learning. It is based on attention mechanisms, enabling the effective handling of long-term relationships in data, making it widely used in applications such as natural language processing and computer vision. See Fig. 6.1.

Turing Test A test of an AI's ability to be indistinguishable (to a human observer) from human intelligence. In Alan Turing's original conception, an AI would be judged by its ability to converse with a human through written text.

References

AIAA (2020), AIAA Institute Position Paper on Digital Twin (December 2020)

J. S. V. Álvarez (2024), The risks and inefficacies of AI systems in military targeting support, September 4, 2024, https://blogs.icrc.org/law-and-policy/2024/09/04/the-risks-and-inefficacies-of-ai-systems-in-military-targeting-support/

C. Anderson (2008), The End of Theory: The Data Deluge Makes the Scientific Method Obsolete, *Wired*, June 23, 2008, https://www.wired.com/2008/06/pb-theory/

K. Appel and W. Haken (1989), **The Four-Color Problem**, American Mathematical Society, 1989

B. Batty (2018), Digital twins. *Environment and Planning B: Urban Analytics and City Science*, 45(5), 817–820.

S. Beiker (2020), Tesla's Use of Digital Twins for Predictive Maintenance and Autonomy. *SAE International*.

I. Bode and I. Bhila (2024), The problem of algorithmic bias in AI-based military decision support systems, September 3, 2024, https://blogs.icrc.org/law-and-policy/2024/09/03/the-problem-of-algorithmic-bias-in-ai-based-military-decision-support-systems/

N. Bostrom (2014), **Superintelligence: Paths, Dangers, Strategies**, Oxford University Press, 2014

F. Caforio, F. Regazzoni, S. Pagani, E. Karabelas, C. Augustin, G. Haase, G. Plank, and A. Quarteroni (2024), Physics-informed neural network estimation of material properties in soft tissue nonlinear biomechanical models, *Computational Mechanics* **75**, 1–27 (2024)

N. Chomsky (1986), **Knowledge of Language**, New York: Praeger, 1986

B. J. Copeland (2024), Alan Turing, *Encyclopedia Britannica*, 23 July 024, https://www.britannica.com/biography/Alan-Turing

J. Corral-Acero, F. Margara, et al. (2020), The 'Digital Twin' to enable the vision of precision cardiology. *European Heart Journal*, 41(48), 2020, 4556–4564.

E. W. Dijkstra (1959), A note on two problems in connection with graphs, *Numerische Mathematik* **1**, 269–271 (1959), https://doi.org/10.1007/BF01386390. S2CID 123284777

M. Fedele, R. Piersanti, F. Regazzoni, M. Salvador, P.C. Africa, M. Bucelli, A. Zingaro, L. Dede' and A. Quarteroni (2023), A comprehensive and biophysically detailed computational model of the electromechanics of the whole human heart, *Computer Methods in Applied Mechanics and Engineering* 410 (2023), https://doi.org/10.1016/j.cma.2023.115983

A. Ferrari and K. Willcox (2024), Digital twins in mechanical and aerospace engineering, *Nature Computational Science*, Vol. 4, No. 3, March 2024, pp. 178–183.

Y. Gao, Y. Xiong, X. Gao, K. Jia, J. Pan, Y. Bi, and H. Wang (2023), Retrieval-augmented generation for large language models: A survey, *arXiv preprint arXiv:2312.10997*.

I. Goodfellow, J. Bengio, and A. Courville (2016), **Deep Learning**, The MIT Press

A. Green and L. Lamby (2023), The supply, demand and characteristics of the AI workforce across OECD countries, OECD Social, *Employment and Migration Working Papers*, No. 287, OECD Publishing, Paris, 2023

R. J. Greenspan and H. A. Dierick. (2004) Am not I a fly like thee? From genes in fruit flies to behavior in humans. *Hum Mol Genet*. 13 Spec No 2:R267-73. https://doi.org/10.1093/hmg/ddh248. PMID: 15358734.

T. C. Hales (2003), A Proof of the Kepler Conjecture, *Discrete & Computational Geometry* **36**(1), 21–69 (2003)

Y. N. Harari (2018), **21 Lessons for the XXI Century**, Spiegel and Grau, 2018

P. E. Hart, N. J. Nilsson, and B. Raphael (1968), A Formal Basis for the Heuristic Determination of Minimum Cost Paths, *IEEE Transactions on Systems Science and Cybernetics* **4**(2), 100–107 (1968)

J. J. Hopfield (1982), Neural networks and physical systems with emergent collective computational abilities, *Proceedings of the National Academy of Sciences* **72**(8), 2554–2558 (1982)

M. I. Jordan and T. M. Mitchell (2015), Machine learning: trends, perspectives, and prospects, *Science* 349.6245, 255–260 (July 2015)

H. Kissinger, E. Schmidt, and D. Huttenlocher (2021), **The Age of AI**, John Murray, London, 2021

M. Klaus (2024), Transcending weapon systems: the ethical challenges of AI in military decision support systems, September 24, 2024, https://blogs.icrc.org/law-and-policy/2024/09/24/transcending-weapon-systems-the-ethical-challenges-of-ai-in-military-decision-support-systems/

J. Landgrebe and B. Smith (2023), **Why Machines Will Never Rule the World**, Routledge, Taylor and Francis Group, 2023

M. Lane, M. Williams and S. Broecke (2023), The impact of AI on the workplace: Main findings from the OECD AI surveys of employers and workers, *OECD Social, Employment and Migration Working Papers*, No. 288, OECD Publishing, Paris, 2034

R. Laubenbacher, B. Mehrad, I. Shmulevich, et al. (2024), Digital twins in medicine. *Nat Comput Sci* 4, 184–191 (2024). https://doi.org/10.1038/s43588-024-00607-6

P. Lewis, E. Perez, A. Piktus, F. Petroni, V. Karpukhin, N. Goyal, and D. Kiela (2020), Retrieval-augmented generation for knowledge-intensive NLP tasks, *Advances in Neural Information Processing Systems* **33**, 9459–9474 (2020)

X. Lu et al. (2024), AI as Humanity's Salieri: Quantifying Linguistic Creativity of Language Models via Systematic Attribution of Machine Text Against Web Test, *OpenReview.net*, 2024

J. Marchant (2020), Powerful Antibiotics Discovered Using AI, *Nature*, 20 February 2020, https://www.nature.com/articles/d41586-020-00018-3

N. Maslej, L. Fattorini, R. Perrault, V. Parli, A. Reuel, E. Brynjolfsson, J. Etchemendy, K. Ligett, T. Lyons, J.Manyika, J.C. Niebles, Y. Shoham, R. Wald, and J. Clark (2024), *The AI Index 2024 Annual Report, AI Index Steering Committee, Institute for Human-Centered AI*, Stanford University, Stanford, CA, April 2024

W. S. McCulloch and W. H. Pitts (1943), A logical calculus of the ideas immanent in nervous activity, *Bulletin of Mathematical Biophysics* **5**, 115–133 (1943)

L. Messeri and M. Crockett (2024), Artificial intelligence and illusions of understanding in scientific research, *Nature* **627** (8002), 49–58 (2024)

M. Minsky (1986), **The Society of Mind**, New York, Simon & Schuster, 1986

L Moura and S Ullrich (2021), The Lean 4 Theorem Prover and Programming Language. In Platzer, André; Sutcliffe, Geoff (eds.). Automated Deduction – CADE 28. *Lecture Notes in Computer Science*, **12699**. Cham: Springer International Publishing. 625–635. https://doi.org/10.1007/978-3-030-79876-5_37

H. P. Moravec (1988), **Mind Children: The Future of Robot and Human Intelligence**, Cambridge, Harvard University Press. 1988

Open AI (2024), GPT-4 Technical Report, arXiv:2303.08774v6, https://doi.org/10.48550/arXiv.2303.08774

Open AI (2024), OpenAI Technical Report 2024, https://arxiv.org/abs/2303.08774

N. Parolini, L. Dede', P. F. Antonietti, G. Ardenghi, E. Miglio, A. Manzoni, A. Pugliese, M. Verani, and A. Quarteroni (2021), SUIHTER: A new mathematical model for COVID-19. Application to the analysis of the second epidemic outbreak in Italy. https://arxiv.org/abs/2101.03369. *Proceedings of the Royal Society A*, 477 (2253) (2021), https://doi.org/10.1098/rspa.2021.0027

N. Parolini, L. Dede', G. Ardenghi and A. Quarteroni (2022), Modelling the COVID-19 and the vaccination campaign in Italy by the SUIHTER model, *Infectious Disease Modelling* 7(2), 45–63 (2022)

A. Petroni (2023), Le ragioni della scienza pura nell'era dell'Intelligenza Artificiale, *SEGNATURE*, November 8, 2023, Accademia Nazionale dei Lincei, Rome

A. Quarteroni (2017), **Numerical Models of Differential Problems**, 3rd edition, Springer Series MS&A, Vol 16, 2017

A. Quarteroni (2022), **Modelling Reality with Mathematics**, Springer, 2022

A. Quarteroni, L. Dede', A. Manzoni and C. Vergara (2019), **Mathematical Modelling of the Human Cardiovascular System. Data, Numerical Approximation, Clinical Applications**, Cambridge University Press, 2019

A. Quarteroni, L. Dede' and F. Regazzoni (2022), Modeling the cardiac electromechanical function: a mathematical journey, *Bulletin of the American Mathematical Society* **59**(3), 371–403 (2022)

A. Quarteroni, P. Gervasio, and F. Regazzoni (2025), Combining physics–based and data–driven models: advancing the frontiers of research with Scientific Machine Learning, *M3AS*, Volume No. 35 (2025), Issue No. 04, pp. 905–1071

A. Quarteroni, F. Saleri and P. Gervasio (2014), **Scientific Computing with MATLAB and Octave**, Springer, 4th edition, 2014

M. Raissi, P. Perdikaris, and G. E. Karniadakis (2021), Physics-informed neural networks: a deep learning framework for solving forward and inverse problems involving nonlinear partial differential equations, *Journal of Computational Physics* 378, 686–707 (2021)

F. Regazzoni, L. Dede' and A. Quarteroni (2020), Machine learning of multiscale active force generation models for the efficient simulation of cardiac electromechanics, *Computer Methods in Applied Mechanics and Engineering* 370 (2020), https://doi.org/10.1016/j.cma.2020.113268

F. Regazzoni, S. Pagani and A. Quarteroni (2022), Universal Solution Manifold Networks (USM-Nets): non-intrusive mesh-free surrogate models for problems in variable domains, *Journal of Biomechanical Engineering* 144.12 (2022): 121004

D. Reinsel, J. Gantz, and J. Rydning (2017), Data Age 2025, *IDC White Paper*, April 2017

F. Rosenblatt (1958), The perceptron: A probabilistic model for information storage and organization in the brain, *Psychological Review* **65**(6), 386–408 (1958)

M. Roser (2023), AI timelines: What do experts in artificial intelligence expect for the future? Published online at *OurWorldinData.org*, retrieved from: https://ourworldindata.org/ai-timelines

J. Searle (2010), **Minds, Brains, and Programs**, The Behavioral and Brain Sciences, Cambridge University Press, 2010

B. Siciliano (2019), Robots are with us, within us and among us, DIID #67, *Design and Technologies – Design, robotics and machines in the post-human age*, 2019

Singapore (2021) **Singapore Government**: *Virtual Singapore: A Digital Twin for the City*, 2021, retrieved from www.nrf.gov.sg

S. Singh (1997), **Fermat's Enigma: The Epic Quest to Solve the World's Greatest Mathematical Problem,** Walker & Co., 1997

R. Salakhutdinov and G. E. Hinton (2009), Deep Boltzman Machines, *Proceedings of the 12th International Conference on Artificial Intelligence and Statistics (AISTATS), 2009, Clearwater Beach, Florida, USA*, Volume 5 of JMLR:W&CP5

G. G. Szapiro (2003), **Kepler's Conjecture**, John Wiley and Sons, 2003

J. Togelius (2024), **Artificial General Intelligence**, MIT Press, 2024

M. Viceconti, A. Henney, and E. Morley-Fletcher (2016), In silico clinical trials: How computer simulation will transform the biomedical industry. *International Journal of Clinical Trials*, 3(2), 2016, 37–46.

A Wiles (1995), Modular Elliptic Curves and Fermat's Last Theorem, *Annals of Mathematics* **141**(3), 443–551(1995)

K. Willcox and B. Segundo (2024), The role of computational science in digital twins, *Nature Computational Science*, Vol. 4, No. 3, March 2024, pp. 147–149.

R. Wilson (2014), ***Four Colors Suffice: How the Map Problem Was Solved***, Princeton Science Library, 2014

D. Zhang et al. (2022), The AI Index 2022 Annual Report, *AI Index Steering Committee, Stanford University HAI*

W. Zhou and A. R. Greipl (2024), Artificial intelligence in military decision-making: supporting humans, not replacing them, August 29, 2024, https://blogs.icrc.org/law-and-policy/2024/08/29/artificial-intelligence-in-military-decision-making-supporting-humans-not-replacing-them/

GPSR Compliance
The European Union's (EU) General Product Safety Regulation (GPSR) is a set of rules that requires consumer products to be safe and our obligations to ensure this.

If you have any concerns about our products, you can contact us on

ProductSafety@springernature.com

In case Publisher is established outside the EU, the EU authorized representative is:

Springer Nature Customer Service Center GmbH
Europaplatz 3
69115 Heidelberg, Germany

www.ingramcontent.com/pod-product-compliance
Lightning Source LLC
LaVergne TN
LVHW010342260326
834688LV00036B/835